Contents

Part 1 Valuation Mathematics

Part 2 The Income Approach

Part 3 Investment Property

Preface to the Fifth Edition

The Income Approach to Property Valuation was first published in 1979 by Routledge and Kegan Paul. At that time we suggested that a lack of basic mathematical ability often prevented valuers from understanding modern or alternative approaches to valuation and, while not deriding the traditional market methods used by previous generations, we hoped that after reading this book trainees, the newly qualified and seasoned valuers would be in a better position to understand the alternative approaches to property valuation.

Much has changed since then and yet in many respects the basic methodology of the subject remains unaltered. The property market has passed through further cycles of boom and slump, has recently aptly illustrated the importance of property as a single property investment, or as part of a balanced portfolio, at a time when other investments, notably in the equity share market, have underperformed. Calculations have been simplified through the rapid growth of IT and enhanced industry standard software. The valuation tasks have become more complex with clients requiring more explicit cash flows, true equivalent yields and 'what if analysis' pushing the profession toward a more evidence based approach to market appraisal.

More emphasis is now placed by the client on the links between performance of property as an investment to the changing social and economic forces, both nationally and internationally. There is an expectation that the valuer will access relevant sources of information and knowledge and will act as an expert advisor on the property market and not simply be a provider of opinions of market value.

Nevertheless understanding the basic concepts of investment and the tools for interpreting market prices and assessing market value remain fundamental to the growing global role of the Chartered Valuation Surveyor in this changing market.

Our aim in this the fifth edition remains consistent with the first edition namely to provide the reader with a clear explanation of how the valuer analyses market rents and sale prices to derive the market evidence to support an opinion of market value; to explain the investment method of valuation and its primary applications simply and unambiguously; to embed the investment method within a thorough appreciation of compound interest, the meaning of the time value of money and the general principles of investment arithmetic; to illustrate how specific legal factors can impact on market value when they interfere with market forces; and to provide a critical view of what the market and the profession might consider to be the 'right' methodology in today's market place.

As far as the latter point is concerned we have made a positive decision in this edition to embrace the single rate approach to the valuation of leasehold investments in property and to regard the dual rate approach as an outdated and unsupportable tool of market analysis and hence an unsupportable method for assessing the market value of leaseholds. We have retained much of the previous material relating to the dual rate approach for the benefit of educators but sincerely hope the market will continue with the current trend away from its usage in the leasehold investment market.

As this book is mainly about principles we have, as previously, left a number of examples unadjusted for inflation; if the concepts can be grasped with small numbers then the trainee may be less concerned when faced with the reality of larger multi million pound valuations. This edition contains a number of spreadsheets, these have been developed for educational purposes only.

The world of publishing has also changed since the first edition. Four successful editions with regular reprints were published by International Thomson Business Press, the successor to RKP. Their business has changed and so have we, this edition becomes the first to be published by the Estates Gazette.

As always we end with a reminder that anyone who gives professional advice or makes any investment decision based on any part of this book does so entirely at their own risk. The opinions expressed in this book, unless indicated otherwise, are the authors' own opinions and do not necessarily represent the views or teaching philosophy of either The University of Reading or Sheffield Hallam University.

Andrew, David and Nick
April 2006.

Acknowledgements

We are grateful to Malcolm Martin BSc FRICS FNAEA of Harvey and Wheeler, Chartered Surveyors for reviewing the section on residential statutory valuation in Chapter 10.

We continue to acknowledge the support given by Peter Byrne and David Jenkins in getting us started with the first edition.

Our thanks to Circle Systems for permission to make use of the software illustrated in Chapter 11 and Appendix C.

Introduction

Cairncross in his *Introduction to Economics* expresses his view that 'economics is really not so much about money as about some things which are implied in the use of money. Three of these-exchange, scarcity and choice-are of special importance.' Legal interests in land and buildings, which for our purposes will be known as property, are exchanged for money and are scarce resources. Those individuals fortunate to have surplus money have to make a choice between its alternative uses. If they choose to buy property they will have rejected the purchase of many other goods, services, alternative investments such as stocks and shares, or of simply leaving it in a savings account or doing nothing. Having chosen to use surplus money to purchase property they will then have to make a choice between different properties. Individuals investing in pension schemes, endowment or with profits life assurance are entrusting their money to others to make similar choices on their behalf.

Valuation is the vocational discipline within economics that attempts to aid that choice in terms of the value of property and the returns available from property. Value in this context can mean the market value in exchange for property rights, as well as value to a particular person or institution with known objectives, currently referred to as an appraisal or assessment of worth or investment value.

Property is purchased for use and occupation or as an investment. In both cases the purchaser measures the expected returns or benefits to be received from the property against the cost outlay. The valuer's task is to express these benefits in money terms and to interpret the relationship between costs and benefits as a rate of return, thus allowing the investor to make a choice between alternative investment opportunities. Current proposals to offer property investment opportunities through Real Estate Investment Trusts (REITs) will strengthen this demand.

Since 1945 the property and construction industries have grown in importance and owning property has been indiscriminately considered to be a 'safe' investment. The growth in pension schemes, life funds, property unit trusts and the like has completed the transition of the property market into a multi-million pound industry. As a result there has been a growth in demand for property to be valued to establish market value and revalued for investors for portfolio and asset management purposes.

Property as an investment is different to other forms of investment. The most obvious difference is its fixed location geographically and hence the importance of the quality of that location for the land's current or alternative uses as determined by its general and special accessibility and its interrelationship with other competing and complementary buildings, locations and land uses. Once developed, the quality of the investment is influenced by the quality of the permitted planning use and the quality of the physical improvements (buildings) on the site. In addition and essential to the assessment of exchange value is the quality of the legal title — is it freehold or leasehold? The owner of a freehold title effectively owns all the land described in the title deeds in perpetuity, including everything below it to the centre of the earth and everything above. Freehold rights may be restricted by covenants in the title and/or by the rights of others such as rights of way. A leaseholder's rights are limited in time, the length of the lease, and by the terms and conditions (covenants) agreed between landlord and tenant and written into the lease or implied or imposed by law or statute. The market value of a tenanted feehold property will also be affected by the quality of the tenant in terms of their covenant strength and the quality of the lease in terms of the appropriateness of the lease conditions.

To be competent the valuer must be aware of all the factors and forces that make a market and which are interpreted by buyers, sellers and market makers in their assessment of market price. In an active market where many similar properties with similar characteristics and qualities are being exchanged, a valuer will, with experience, be able to measure exchange value by comparing that which is to be valued with that which has just been sold. This direct or comparative method of valuation is used almost exclusively for the valuation of vacant possession freehold residential property. Differences in age, condition, accommodation and location can all, within reason, be reflected by the valuer in the assessment of value. Differences in size can sometimes be overcome by adopting a unit of comparison such as price per hectare or rent per square metre.

The more problematic properties are those for which there is no ready market, those which display special or unique characteristics, those which do not fully utilise the potential of their location and are therefore ripe for development, redevelopment, or refurbishment, those that are tenanted and are sold as investments at prices reflecting their income generating potential; and leasehold properties.

For each of these broad categories of property, valuers have developed methods of valuation that they feel most accurately reflect the market's behavioural attitude and which may therefore be considered to be rational methods.

In the case of special properties such as oil refineries, glassworks, hospitals and schools, the usual valuation method is the cost or contractor's method. It is the valuer's method of last resort and is based on the supposition that no purchaser would pay more for an existing property than the sum of the cost of buying a similar site and constructing a similar building with similar utility written down to reflect the physical, functional, environmental and locational obsolescence of the actual building.

Properties with latent development value are valued using the residual or development (developer's) method. The logic here is that the value of the property (site) in its current state must equal the value of the property in its developed or redeveloped state less all the costs of development including profit but excluding the land. In those cases where the residual sum exceeds the existing use value then in theory the property will be released for that higher and better use.

All property that is income producing or is capable of producing an income in the form of rent, and for which there is both an active tenant market and an active investment market, will be valued by the market's indirect method of comparison. This is known as the investment method of valuation or the income approach to property valuation and is the principal method considered in this book.

The income approach and the income-based residual warrant special attention if only because they are the valuer's main weapon in the valuation of the most complex and highly priced investment properties.

The unique characteristics of property make property investment valuation more complex an art and science than that exercised by brokers and market makers in the market for stocks and shares. As stocks, shares and property are the main investments available there is bound to be some similarity between the pricing (valuation) methods used in the various markets and some relationship between the investment opportunities offered by each. A basic market measure is the investment yield or rate of return. The assessment of the rate of return allows or permits comparison to be made between investments in each market and between different investments in different markets. There is a complex interrelationship of yields and patterns of yields within the whole investment market. In turn these yields reflect market perceptions of risk and become a key to pricing and valuation methods. Understanding market relationships and methods can only follow from an understanding of investment arithmetic.

Part One of this book therefore considers the basic investment arithmetic used by valuers. Part Two applies that knowledge to the valuation of income-producing property. Part Three considers some of the techniques used in the field of risk analysis. Throughout it is assumed that the reader has some knowledge of who buys property, why they buy it and what alternative investment opportunities there are, and also that they will have some knowledge of the nature of property as an investment. The reader should have some awareness of the social, economic and political factors that influence the market for and the value of the property.

Our purpose in writing this book was aptly summarised, coincidentally, by Robert Chartham in *Sex Manners for Advanced Lovers*, which we misquote with apologies:

> What we are trying to aim at is to put forward suggestions of techniques which may not have occurred to some who have already transformed themselves into highly proficient (seasoned) valuers, in the hope that they will be encouraged to try them out to discover for themselves whether or not they are of any help to them, their partners or their clients.

Part 1

Valuation Mathematics

Single Rate Tables

This book explores the process of property valuation with particular reference to property which is bought or sold as an investment. In order to be able to value an investment property a valuer must understand how the benefits to be enjoyed from the ownership of a freehold or leasehold interest in property can be expressed in terms of present value.

To do this a valuer must have a working knowledge of the mathematics of finance and the theory of compounding and discounting as it relates to savings and investments.

This chapter explores the mathematics behind the six investment functions of £1 and illustrates their use in the practice of property valuation.

The six functions of £1

The six basic functions of financial mathematics are:

- **The Amount of £1 (A)** another name for compound interest; that is, the future worth of £1 invested today allowing for compound interest at a given rate.
- **The Amount of £1 per annum (A £1 pa)** that is, the future worth of £1 invested at the end of each year accruing compound interest at a given rate.
- **The Annual Sinking Fund (ASF)** that is, the fraction of £1 which must be invested at regular intervals to produce £1 at a given point in the future with compound interest accruing at a given rate.
- **The Present Value of £1 (PV£1)** that is, the present value of £1 to be received in the future, discounted over a given period at a given rate.
- **The Present Value of £1 per annum (PV £1 pa)** that is, the present value of a series of payments of £1 due annually for a given period of time, discounted at a given rate. This function is also known by valuers as the Years' Purchase (YP) single rate.
- **The Annuity £1 Will Purchase (AWP)** that is, the amount of money that will be paid back at the end of each year in return for £1 invested today for a given number of years at a given rate.

There are several published valuation tables such as *Parrys*, published by the Estates Gazette, which show the value of these and other functions for a range of interest rates and years. Some of these tables compute income as being received or invested quarterly in advance, as with many actual investments, however, in order to simplify this introductory chapter it is assumed initially that income is received or invested at the beginning or end of each year in one instalment.

The following standard notation is used throughout:

i =   interest, expressed as a decimal
          ie 5% = 5/100 = 0.05
n =   number of time periods, usually the number of years.

While reference is made throughout to pounds (GBP £) the principles apply to any currency and the Euro or US dollar or any other currency can be substituted.

Percentages are used frequently by valuers but as frequently their calculation is forgotten from school days. One per cent (1%) is one hundredth part of the whole (1/100th). If a gift of £1,000 is to be divided equally between 10 people then it has to be divided in to ten parts. Each person will receive 1/10th or 10% [1/10th being (1/10 × 100) = 10%]. So 1/10th or 10% of £1,000 is [(£1,000/100) × 10] which is £100.

The amount of £1

The amount of £1 is simply another term for compound interest. Consider the building society passbook set out below where the interest rate for deposits is at 10%pa.

Date	Description	Deposits	Withdrawals	Interest	Balance
01/01/02	Cash	£100			
31/12/02	Interest			£10.00	£110.00
31/12/03	Interest			£11.00	£121.00
31/12/04	Interest			£12.10	£133.10
31/12/05	Interest			£13.31	£146.41

This table shows that the interest accumulates on both the original £100 invested and on the interest added to it.

The formula to express this states that if £1 is invested for 1 year at i interest, then at the end of one year it will have accumulated to $(1 + i)$. So at 5% this will read as $(1 + 0.05)$ and the initial £1 with interest added becomes £1.05.

At the end of the second year $(1 + i)$ will have earned interest at i, so at the end of the second year it will have accumulated to:

$(1 + i) + i(1 + i)$ which can be expressed as

$(1 + i)^2$

At the end of n years the accumulated sum will be

$(1 + i)^n$

Remember that interest is expressed as a decimal (*i*) or (*r*), eg 10% = 10 divided by 100 = 0.10 and that while *n* is normally the number of years of interest accumulation the formula or equation is a general one for compound interest. So if interest is added monthly *n* for one year becomes 12 and *i* will be the monthly rate of interest.

Example 1.1

Calculate the amount of £1 after 4 years at 10%.

$$A \ = \ (1+i)^n \qquad\qquad i = 0.10; \ n = 4$$

$$A \ = \ (1.1)^4$$

$$= \ (1.1) \times (1.1) \times (1.1) \times (1.1)$$

$$= \ 1.4641$$

The calculation shows that £1 will accumulate to £1.4641 (£1.47 to the nearest pounds and pence) after 4 years at 10% compound interest rate. Notice that if the figure produced by the formula in Example 1.1 (1.4641) is multiplied by 100 the figure is the same as that shown in the building society passbook.

The amount of £1 pa

This function is used to calculate the future value, with compound interest, of a series of payments made at regular intervals, normally each year hence the Latin per annum. Many investments follow this pattern. Consider for example another building society passbook set out below where this time the investor deposits £100 at the end of each year and interest is added at 10% pa.

Date	Description	Deposits	Withdrawals	Interest	Balance
31/12/01	Cash	£100			
31/12/02	Interest			£10.00	£110.00
31/12/02	Cash	£100			£210.00
31/12/03	Interest			£21.00	£231.00
31/12/03	Cash	£100			£331.00
31/12/04	Interest			£33.10	£364.10
31/12/04	Cash	£100			£464.10
31/12/05	Interest			£46.41	£510.51
31/12/05	Cash	£100			£610.51

The amount of £1 pa deals with this type of investment pattern. It indicates the amount to which £1, invested at the end of each year, will accumulate at *i* interest after *n* years.

The table is simply a summation of a series of amounts of £1. If each £1 is invested at the end of the year, the nth £1 will be invested at the end of the nth year and will thus earn no interest. See last cash entry above for 31/12/05.

Each preceding £1 will earn interest for an increasing number of years:

the $(n - 1)$ £1 will have accumulated for 1 year, and will be worth $(1 + i)$
the $(n - 2)$ £1 will have accumulated for two years, and will be worth $(1 + i)^2$
the first £1 invested at the end of the first year will be worth $(1 + i)^{n-1}$.

This series of calculations when added together is expressed as:

$$1 + (1 + i) + (1 + i)^2 \ldots (1 + i)^{n-1}$$

This is a geometric progression and when summed it can be expressed as:

$$\frac{(1 + i)^n - 1}{i} = \text{formula for the amount of £1 pa (A £1 pa)}$$

Example 1.2

Calculate the amount of £1 pa for 5 years at 10%.

$$\text{A £1 pa} = \frac{(1 + i)^n - 1}{i} \qquad i = 0.10; \ n = 5$$

$$\text{A £1 pa} = \frac{(1.10)^5 - 1}{i} = \frac{1.61051 - 1}{0.10}$$

$$= \frac{0.61051}{0.10} = 6.1051$$

Multiplying this figure by 100 to calculate the sum that £100 invested at the end of each year will accumulate to after five years produces the same figure as in the building society passbook.

Example 1.3

If Mr A invests £60 in a building society at the end of each year, and at the end of 20 years has £4323, at what rate of interest has this sum accumulated?

Annual sum invested	£60
A £1 pa for 20 years at i%	x
Capital Value (CV) of Investment at the end of 20 years is	£4,323

Rephrasing as a simple equation:

£60x =   £4323 ie £60 multiplied by an unknown number x will produce a sum of £4323 and so:

$$x \quad = \frac{£4,323}{£60} = 72.05$$

If the reader refers to *Parrys* Amount of £1 pa tables, it will be seen that 72.05 is the value for 20 years at 12% which is the rate of compound interest at which this regular investment has accumulated.

Annual sinking fund

Property investors may require to make provision from today for some future expenditure. This may be for some form of maintenance, for example, a building may require a new roof or service roads may require resurfacing.

Such obligations may be passed on to the tenant in the form of service charges, payable in addition to rent, which include the provision of a fund to be used for major works in the future. However, in some cases the landlord may be responsible for major repairs which cannot be recovered from the tenant. Such obligations should be reflected in the purchase price of the investment. Such expenditure can be budgeted for by investing a lump sum today or by investing a regular sum from today, both of which will grow due to compound interest. A property owner may need to know how much should be invested now or on an annual basis.

The first step is to estimate the probable amount needed at an expected future date.

The investor could then either invest a lump sum immediately which with the accumulation of interest will meet the estimated outlay when it arises. This can be calculated using the Present Value of £1 table (see later). But it may be more effective to set aside part of the income received from the investment (ie rent) regularly in an account known as a sinking fund, which is planned to accumulate to the required sum by the date in the future when the expenditure is required.

This is similar to the amount of £1 pa described above. Example 1.4 demonstrates how the amount of £1 pa tables can be used to calculate the sum required to be set aside in a sinking fund to meet a known future expense.

Example 1.4

An investor is considering the purchase of a small shop in which the window frames have begun to rot. It is estimated that in 4 years time they will require complete replacement at a cost of £1,850. The shop produces a net income of £7,500 pa.

How much of this income should be set aside each year to meet the expense assuming the money is invested with a guaranteed fixed return of 7% per year?

Amount to be set aside	x
A £1 pa for 4 years at 7%	4.44
Cost	£1,850

£1,850   = 4.44x

£1,850/4.44   = x

x   = £416.67 pa. This annual sum is the sinking fund, or ASF.

The use of annual sinking fund tables enables this sum to be calculated more easily.

Sum required	£1,850
× by ASF to replace £1 in 4 years at 7%	0.22523
ASF	£416.67 pa

It can be seen that this calculation performs the function of the amount of £1 pa in reverse. In Example 1.4 the amount to be set aside was found by dividing the capital sum required by the amount of £1 pa.

In the above, the annual sum was found by multiplying the sum required by the annual sinking fund (ASF) to replace £1 at the end of four years.

Therefore the ASF table is the reciprocal of the amount of £1 pa:

$$A\ £1\ pa\ =\ \frac{(1+i)^n-1}{i}$$

$$\text{The amount of £1 (A)} \ =\ (1+i)^n$$

$$\text{Therefore: A £1 pa}\ =\ \frac{A-1}{i}$$

$$\text{ASF}\ =\ \frac{i}{(1+i)^n-1}\ =\ \frac{i}{A-1}$$

Example 1.5

Calculate the ASF to accumulate to £1 after 4 years at 7%

$$\text{ASF}\ =\ \frac{i}{(1+i)^n-1}\ =\ \frac{0.07}{(1.07)^4-1}$$

$$=\ \frac{0.07}{1.31079-1}\ =\ 0.22523$$

The present value of £1

The first three functions of £1 have shown how any sum invested today will be worth more at some future date due to the accumulation of compound interest. This means that £1 receivable in the future cannot be worth the same as £1 at the present time. What it is worth will be the sum that could be invested now to accumulate to £1 at a given future date. This sum will obviously depend upon the length of time over which it is invested and the rate of interest it attracts.

If £1 were invested now at a rate of interest i for n years, then at the end of the period it would be worth $(1 + i)^n$.

If £x were to be invested now at i for n years and it accumulates to £1:

Then £x$(1 + i)n$ $=$ £1

And £x $=$ £1 $\times$ $\dfrac{1}{(1 + i)^n}$

This is the formula for the present value of £1 (PV), and it is the reciprocal of the amount of £1.

$$PV = \frac{1}{(1 + i)^n} = \frac{1}{A}$$

Proof :

PV £1 in 7 years at 10% $=$ 0.51316
A £1 in 7 years at 10% $=$ 1.9487
0.51316 $\times$ 1.9487 $=$ 1.00

Example 1.6

If Mr X requires a rate of return of 10%, how much would you advise him to pay for the right to receive £200 in 5 years time?

$$PV = \frac{1}{(1 + i)^n} = \frac{1}{(1.10)^5} = \frac{1}{1.6105}$$

$$= 0.6209 \times £200 = £124.18$$

This means that if £124.18 is invested now and earns interest at 10% each year then it would accumulate with compound interest to £200 in 5 years time.

The present value of £1 pa (or years' purchase)

The amount of £1 pa was seen to be the summation of a series of amounts of £1. Similarly, the present value of £1 pa is the summation of a series of present values of £1. It is the present value of the right to receive £1 at the end of each year for *n* years at *i*.

The present value of £1 receivable in one year is:

$$\frac{1}{(1 + i)}$$

in two years it is:

$$\frac{1}{(1 + i)^2}$$

and so the series reads:

$$\frac{1}{(1 + i)} \quad + \quad \frac{1}{(1 + i)^2} \quad + \quad \frac{1}{(1 + i)^3} \quad \ldots \quad \frac{1}{(1 + i)^n}$$

This is a further geometric progression which when summated can be expressed as:

$$\frac{1 - \dfrac{1}{(1 + i)^n}}{i}$$

This is the formula for the PV of £1 pa, and if:

$$\frac{1}{(1 + i)^n} = PV$$

Then the PV £1 pa can be simplified to $\dfrac{1 - PV}{i}$

Example 1.7

Calculate the present value of £1 pa at 5% for 20 years given that the present value of £1 in 20 years at 5% is 0.3769

$$\text{PV £1 pa} \quad = \quad \frac{1 - PV}{i}$$

$$= \quad \frac{1 - 0.3769}{0.05} \quad = \quad \frac{0.6231}{0.05}$$

$$= \quad 12.462$$

Example 1.8

How much should A pay for the right to receive an income of £675 for 64 years if he/she requires a 12% return on the investment?

Income per year	£675
× PV £1 pa for 64 years at 12%	8.3274
Capital Value (CV)	£5,621

The present value of £1 pa is usually referred to by UK property valuers as the 'years' purchase'. The *Oxford English Dictionary* gives a date of 1584 for the first use of this phrase 'at so many years' purchase' used in stating the price of land in relation to the annual rent in perpetuity. This term is sometimes confusing as it does not relate to the other investment terms. However, the terms are interchangeable and both will be used.

Obviously the PV £1 pa will increase each year to reflect the additional receipt of £1. However each additional receipt is discounted for one more year and will be worth less following the present value rule established above. The PV £1 pa in fact approaches a maximum value at infinity. However, as the example below shows, in fact the increase in PV £1 pa becomes very small after 60 years and is customarily assumed for the purpose of property valuation to reach its maximum value after 100 years. In valuation terminology this is referred to as 'perpetuity'.

Example 1.9

In the formula $\dfrac{1 - PV}{i}$ what happens to PV as the time period increases?

What effect does this have on the YP figure ?

Table 1.1

Years	PV at 10%	PV £1 pa at 10%	Years	PV at 10%	PV £1 pa at 10%
1	0.90909	0.90909	90	0.0001882	9.998
2	0.82645	1.736	91	0.0001711	9.998
3	0.75131	2.487	92	0.0001556	9.998
4	0.68301	3.170	93	0.0001414	9.999
5	0.62092	3.791	94	0.0001286	9.999
6	0.56447	4.355	95	0.0001169	9.999
7	0.51316	4.868	96	0.0001062	9.999
8	0.46651	5.335	97	0.0000966	9.999
9	0.42410	5.759	98	0.0000878	9.999
10	0.38554	6.145	99	0.0000798	9.999
			100	0.0000726	9.999
			Perp.		10.000

From Table 1.1 two facts are clear, the PV decreases over time and the YP (PV £1 pa) increases over time. In addition it can be seen that the YP (PV £1 pa) is the accumulation of the PVs.

As n approaches perpetuity the PV tends towards 0; the present value of £1 to be received such a long time in the future is reduced to virtually nothing.

Therefore if PV tends to 0 at perpetuity and if n is infinite then given that:

$$YP = \frac{1 - PV}{i}$$

YP in perpetuity will tend to: $\dfrac{1 - 0}{i}$

The formula for YP in perpetuity (YP perp) is therefore:

$$\frac{1}{i}$$

at a rate of 10% the YP perp $= \dfrac{1}{0.10} = 10$

Example 1.10

A freehold property produces a net income (annual net rent) of £1,500 a year. If an investor requires a return of 8%, what price should be paid? The income is perpetual so a YP in perpetuity should be used.

Income	£1,500	
YP perp at 8%	12.5	$(1/0.08 = 12.5)$
Capital Value	£18,750	

Years' purchase of a reversion to a perpetuity (perpetual income)

A further common valuation application of the present value of £1 pa is known as 'the years' purchase of a reversion to a perpetuity'. It shows the present value of the right to receive a perpetual income starting at a future date and is found by multiplying YP in Perpetuity by the Present Value of £1 at the same rate of interest for the period of time that has to pass before the perpetual income is due to commence. It is useful to property valuers as often property will be assumed to revert to a perpetual higher income after an initial period of time such as following a rent review or renewal of a lease after a period of under renting.

$$\text{YP rev perp} \quad = \quad \text{YP in perp} \times \text{PV £1}$$

$$= \quad \frac{1}{i} \times \frac{1}{(1 + i)^n}$$

this can be simplified to $\dfrac{1}{iA}$

Example 1.11

Calculate in two ways the present value of a perpetual income of £600 pa beginning in 7 years' time using a discount rate of 12%.

Income	£600
YP perp at 12%	8.33
Capital Value (CV)	£5,000
PV £1 in 7 years at 12%	0.45235
CV	£2,262

or:

$$\text{YP rev perp} = \quad \frac{1}{iA} = \frac{1}{i(1+i)^7} = \frac{1}{0.12(1.12)^7} = \frac{1}{0.12(2.2107)} = \frac{1}{0.265284} = 3.7965$$

Income	£600
YP rev perp in 7 years at 12%	3.7695
CV	£2,262

The answer can also be found by deducting the YP for 7 years at 12% from the YP in perpetuity at 12% 8.3334 – 4.5638 =3.7695.

The use of this present value or discounting technique to assess the price to be paid for an investment or the value of an income-producing property ensures the correct relationship between future benefits and present worth; namely that the investor will obtain both a return on capital and a return of capital at the target rate or market-derived rate of interest used in the calculation.

This last point is important and can be missed when PVs are added together to produce the PV of £1 pa and labelled 'Years' Purchase' by the property valuation profession.

Example 1.12

An investor is offered five separate investment opportunities on five separate occasions each will produce a certain cash benefit of £10,000, the first in exactly one years time and the other four at subsequent yearly intervals. The investor is seeking a 10% return from his money. What price should be paid for each investment?

	Year 1	Year 2	Year 3	Year 4	Year 5
Benefit	£10,000	£10,000	£10,000	£10,000	£10,000
PV £1 at 10%	0.9090	0.8264	0.7513	0.6830	0.6209
Present Value (investment price today)	£9,090	£8,264	£7,513	£6,830	£6,209
Amount of £1 at 10%	1.1000	1.2100	1.3310	1.4641	1.6105
	£10,000	£10,000	£10,000	£10,000	£10,000

The figures of present value – £9090, £8264, £7513, £6830, £6209 – show the individual prices to be paid today for each investment in order that the investor can achieve a 10% return on capital invested and obtain the return of the capital invested.

In other words the investor is exchanging a sum of money today for a known future sum which will be equal to the sum of money today plus the interest forgone if the capital had been invested elsewhere at the same rate of interest. In each case the receipt of £10,000 at the due date returns the respective capital sum or price paid and the difference between the price today and the £10,000 is equivalent to the 10% annual compound interest that would have been earned on the purchase sums if they had been saved at the investors 10% required rate of return.

Example 1.13

If an investor is offered a single investment which generates a 'certain', that is a known, cash benefit of £10,000 at the end of each year for the next five years and wishes to achieve a 10% return, what price should they pay today?

The answer can be found by adding the five sums together eg £9,090 + £8,264 + £7,513 + £6,830 + £6,209 which equals a sum of £37,908. But it is quicker to multiply the £10,000 a year by the PV £1 pa for 5 years at 10%.

£10,000 × 3.7908 = £37,908

Again the investor must achieve a return of capital, (the initial £37,908) and a return of 10% on the capital (that is the interest forgone). The proof is shown in the table below:

Year	Capital Outstanding	Return @ 10%	Income	Return of Capital (Balance)
1	37,908.00	3,790.80	10,000	6,209.20
2	31,698.80	3,169.88	10,000	6,830.12
3	24,868.68	2,486.87	10,000	7,513.13
4	17,355.55	1,735.55	10,000	8,264.45
5	9,091.10	909.11	10,000	9,090.89

Note: rounding errors due to calculation to 2 decimal places only

Annuity £1 will purchase

This function shows the amount that will be paid back at the end of each year for *n* years at *i*, in return for £1 invested. It calculates what is known as the annuity that £1 will purchase. It can be used to calculate the annual equivalent of a capital sum although valuers tend to prefer to divide a capital sum by the YP (PV of £1 pa) to calculate an annual equivalent.

An annuity entitles the investor to a series of equal annual sums. These sums may be perpetual or for a limited number of years.

When money is invested in a building society, interest accumulates on the principal which remains in the account. When an annuity is purchased, however, the initial purchase price is lost forever to the investor. The return from building society accounts is all annual interest on capital. The return from an annuity represents partly interest on capital but also partly the purchase price being returned bit by bit to the investor.

These constituent parts will be referred to as: **return on capital (interest) and return of capital**.

In an annuity the original capital outlay must be returned by the end of the investment. If not, how could the rate of interest earned by an annuity investment be compared with the rate of interest earned in a building society account?

The amount of an annuity will depend upon three factors: the purchase price, *i* and *n*.

Example 1.14

What annuity will £50 purchase for 5 years if a 10% yield is required by the purchaser?

Purchase sum	£50
Annuity £1 wp for 5 years at 10%	0.2638
Annuity	£13.19

The £13.19 has two constituent parts. The return on capital is 10% of the outlay, ie £5 pa. This leaves £8.19 extra. This is the return of capital. But 5 × £8.19 does not return £50 because each payment is in the nature of a sinking fund instalment and assumed to be earning interest, in this case at 10%.

Proof :

ASF	£8.19
A £1 pa for 5 years at 10%	6.1051
CV	£50

Example 1.14 shows that the return *on* capital and return *of* capital are both achieved, as a sinking fund is inherent in the annuity.

It can be noted that if the return on capital = *i* and the return of capital is SF then the formula for the annuity £1 wp must be *i* + SF, where *i* is the rate of interest required and SF is the annual sinking fund required to replace the capital outlay in *n* years at *i*%.

Example 1.15

Calculate the annuity £1 will purchase for 10 years @ 10%, if £6.1446 will purchase £1 at the end of each year for 10 years.

£1 will purchase $\dfrac{1}{6.1466}$ = £ 0.1627

Therefore the annuity £1 will purchase for 10 years at 10% is £0.1627.

The annuity factor can also be calculated from the formula i + SF. As the sinking fund factor is the annual factor of £1 needed to build up the return of capital, this is known technically as to amortise £1, so the total partial payment required for recovery of capital and for interest on capital must be the amortisation factor plus the interest rate.

Therefore the annuity £1 will purchase for 10 years at 10% is:

	i =	0.1000
plus the ASF to replace £1 in 10 years at 10%	=	0.0627
		£0.1627

An annuity generally means a life annuity, a policy issued by a life assurance company where the investor will receive for the rest of his/her life a given annual income in exchange for a given capital sum. The calculations undertaken by the life office's actuaries have to take into account many factors — including life expectancy, which is determined by lifestyle. This is beyond the scope of this book, which is concerned with annuities related to property where the period of time the annuity will be paid is certain. Before proceeding with further consideration and analysis of annuities there are a number of common terms which should be explained.

Annuity terminology

In arrears: This means that the first annual sum will be paid (received) 12 months after the purchase or taking out of the policy. The payments could be weekly, monthly, quarterly or for any period provided they are in arrears.

In advance: This means that the payments are made at the beginning of the week, month, year, etc.

An immediate annuity: The word 'immediate' is used to distinguish a normal annuity from a *deferred* annuity. An immediate annuity is one where the income commences immediately either 'in advance' or 'in arrears', whereas in the case of a deferred annuity capital is exchanged today for an annuity 'in advance' or 'in arrears', the first such payment being deferred for a given period of time longer than a year. In assurance terms one might purchase at the age of 50 a life annuity to begin at the age of 65. This would be a deferred annuity.

The majority of annuity tables are based on an 'in arrears' assumption.

The distinction between *'return on' and 'return of' capital* is important in the case of life annuities because the capital element is held by the Inland Revenue to be the return of the annuitant's capital and therefore is exempt from income tax. In practice the Inland Revenue have had to indicate how the distinction is to be made. Quite clearly a precise distinction is not otherwise possible because of the uncertain nature of the annuitant's life.

The distinction in the case of certain property valuations undertaken on an annuity basis is important, because the Inland Revenue are not allowed to distinguish between 'return on' and 'return of' capital other than for life annuities. Thus tax may be payable on the whole income from the property and the desired return may not be achieved unless this factor is accounted for in the valuation.

Example 1.16a

What annuity will £4,918 purchase over 6 years at 6%?

	£4,918
Annuity £1 wp for 6 years at 6%	0.2033
	£1,000 pa

Example 1.16b

What annuity will £1,000 purchase over 3 years at 10%?

	£1,000
Annuity £1 wp for 3 years at 10%	0.4020
	£402 pa

Prove that the investor recovers his capital in full and earns interest from year to year at 10%.

Year	Capital outstanding	Interest at 10%	Income (annuity)	Return of Capital	
1	£1,000	£100	£402	£302	(£402 – £100)
2	£698 (£1,000 – £302)	£69.80 (10% on £698)	£402	£332.20	(£402 – £69.80)
3	£365.80 (£698 – £332.20)	£36.58 (10% on £365.80)	£402	£365.42	(£402 – £36.58)
				+ £999.62	(error due to rounding)

Example 1.17

What sum would have to be paid today to acquire an annuity of £1,000 for 6 years in arrears at 8% to begin in 2 years time?

The 'in arrears' assumption means that the first £1,000 will be paid in 36 months' time.

Annuity income	£1,000
PV £1 pa for 6 years at 8%	4.622
Cost of immediate annuity	£4,622
Defer 2 years × PV £1 in 2* years at 8%	0.857
	£3,961

* the PV table assumes payment in arrears

An annuity may be payable in advance.

This should cause no difficulty as the PV of £1 pa payable in advance at rate i for n periods is simply £1 plus the PV of £1 per period, payable in arrears at rate i for $(n-1)$ periods.

Alternatively one can multiply the PV £1 pa in arrears for n periods by the factor $(1+i)$, (see pp42–45).

Example 1.18

Calculate the capital cost of an annuity of £500 for 6 years due in advance at 10%.

Annuity income		£500
PV £1 pa for 5 years $(n-1)$ at 10%	3.79	
plus	1.00	
		4.79
		£2,395

or		£500
PV £1 pa for 6 years at 10%	4.3553	
	× 1.1	4.79
		£2,395

or £500 in advance for 6 years is £500 in arrears for 5 years plus an immediate £500:

	£500
PV £1 pa for 5 years at 10%	3.79
	£1,895
	+£500
	£2,395

Annuities may be variable. The present worth of variable income flows could be expressed as

$$V = \frac{I_1}{PV_1} + \frac{I_2}{PV_2} + \frac{I_3}{PV_3} \quad ... \quad \frac{I_n}{PV_n}$$

While it is possible to have a variable annuity changing from year to year it is more common to find the annuity changing at fixed intervals of time.

Example 1.19

How much would it cost today to purchase an annuity of £1,000 for 5 years followed by an annuity of £1,200 for 5 years on a 10% basis ?

Immediate annuity		£1,000	
PV £1 pa for 5 years at 10%		3.79	£3,790
Plus deferred annuity		£1,200	
PV £1 pa for 5 years at 10%	3.79		
PV £1 for 5 years at 10%	0.62	2.3498	£2,820
			£6,610

It will be shown later that this approach is the same as that used by valuers when valuing a variable income flow from a property investment.

In the formula $i + SF$, i will remain the same whatever the length of the annuity. But the value of each year of the annuity will reduce as time increases due to the effect of SF. For a perpetual annuity, SF will be infinitely small, and tends towards 0. A perpetual annuity therefore $= i$.

Example 1.20

What perpetual annuity can be bought for £1,500 if a 12% return is required by the investor?

Invested sum	£1,500
($i = 0.12$)	
Annuity £1 wp at 12% in perp	0.12
	£180 pa

The formula for a perpetual annuity is i; the formula for a Years' Purchase in perpetuity is $1/i$ and is therefore the reciprocal. The annuity £1 will purchase is the reciprocal of the present value of £1 pa and it follows that there must be a second formula for Years' Purchase namely:

$$\frac{1}{i + SF}$$ as the limited term annuity formula is $i + SF$.

This is easily proved.

Annuity £1 wp $= i + SF$:

hence capital $\times (i + SF)$ $=$ Income
and income $\times$ YP $=$ Capital

Therefore

$$YP = \frac{1}{i + SF}$$

Example 1.21

How much should A pay for the right to receive £1 pa for 25 years at 10%?

$$\text{YP} = \frac{1}{i + \text{SF}} = \frac{1}{i + \dfrac{i}{(1+i)^n - 1}} = \frac{1}{0.10 + \dfrac{0.10}{(1+0.10)^{25} - 1}} = \frac{1}{0.10 + 0.010158} = 9.077$$

If the YP or Present Value of £1 pa is 9.077 then A should pay £9.077.
Check

$$\text{YP} = \frac{1 - \text{PV}}{i} = \frac{1 - 0.092296}{0.10} = £9.077$$

The function of SF in this YP formula is exactly the same as that of PV in the original — that is, to reduce the value of the YP figure as n decreases. The Years' Purchase figure must include a sinking fund element, as the valuation of an investment involves equating the outlay with the income and the required yield, so that both a return on capital and a return of capital are received.

Example 1.22

Value an income of £804.21 pa receivable for 3 years at a required yield of 10%. Illustrate how a return on and a return of capital are achieved.

Income	£804.21
YP for 3 years at 10%	2.4869
	£2,000

The return *on* and a return *of* capital can be demonstrated by constructing a table:

Year	Capital outstanding	Interest on capital outstanding	Income	Return of Capital
1	£2,000	£200	£804.21	£604.21 (£804.21 − £200)
2	£1,395.79	£139.58	£804.21	£664.63 (£804.21 − £139.58)
3	£731.16	£73.12	£804.21	£731.09 (£804.21 − £73.12 error due to rounding)

A rate of return is received on outstanding capital (ie that capital which is at risk) only. The table assumes that some capital is returned at the end of every year so that the amount of capital outstanding is reduced year by year. Interest on capital therefore decreases and more of the fixed income is available to return the capital outstanding. The last column shows how the sinking fund accumulates. £604.21 is the first instalment: £664.63 represents the second instalment of £604.21 plus one year's interest at 10%. £731.09 represents the third and final sinking fund instalment of £604.21 plus one year's interest at 10% on £1,268.84 (£664.63 + £604.21).

The three return of capital figures summate to the original outlay of £2,000.

The above type of table is also used to show how a mortgage is repaid.

The use of single rate tables and mortgages

When a property purchaser borrows money by way of a legal mortgage it is usually agreed between the parties that the loan will be repaid in full by a given date in the future. Like anyone else lending money, the building society or bank, known as the mortgagee, will require the capital sum to be repaid and will require interest on any outstanding amounts of the loan until such time as the loan and all interest are recovered. This is comparable to the purchase of an annuity certain. Indeed, conceptually, from the point of view of the mortgagee it is the purchase of an annuity. It follows that the mortgagee will require a return on capital and the return of capital.

The repayment mortgage allows the mortgagor (borrower) to pay back a regular sum each year, often by equal monthly instalments. This sum is made up of interest on the capital owed and the partial return of capital. In the early years of the mortgage most of the payment represents interest on the loan outstanding and only a small amount of capital is repaid. But over the period of the loan, as more and more capital is repaid, the interest element becomes smaller and the capital repaid larger.

For the purpose of explanation, interest is assumed fixed throughout the term of the loan. In practice lenders offer a wide range of mortgage deals but the underlying concept is the same — a requirement that capital leant will be repaid and that interest will be paid on money owed. The rates of interest and pattern of payment can be infinitely variable and often involve lower interest payments during the early years.

The valuer who can understand the concept of the normal repayment mortgage and can solve standard problems that face mortgagors and mortgagees should readily understand most investment valuation problems. The following examples are indicative of such mortgage problems.

Example 1.23

The sum of £10,000 has been borrowed on a repayment mortgage at 6% for 25 years.

(i)   Calculate the annual repayment of interest and capital.
(ii)   Calculate the amount of interest due in the first and tenth years.
(iii)   Calculate the amount of capital repaid in the first and tenth years.

The formula for the annuity £1 will purchase is $i + \text{SF}$.

Its reciprocal, the PV £1 pa is $\dfrac{1}{i + \text{SF}}$ or $\dfrac{1 - \text{PV}}{i}$

The mortgagee is effectively buying an annuity. There is in effect an exchange of £10,000 for an annual sum (annuity) over 25 years at 6%. What is the annual sum?

Mortgage sum	£10,000
Annuity £1 wp for 25 years at 6%	0.07823
Annual Repayment	£782.30

Or £10,000 divided by PV £1 pa for 25 years at 6% which is £10,000 /12.7834 = £782.30.
As interest in the first year is added immediately to capital borrowed, then:

£10,000 × 0.06 = £600 interest due in first year.

As the total to be paid is £782.30 so the amount of capital repaid will be £782.30 – £600 = £182.30 in the first year.
The amount of interest in the tenth year will depend upon the amount of capital outstanding at the beginning of the tenth year, ie after the ninth annual payment.
Although a mortgage calculation is on a single rate basis, as the formula is i + SF one can assume that £182.30 a year is a notional sinking fund accumulating over 9 years at 6%.

	£182.30
Amount of £1 pa for 9 years at 6%	11.4913
Capital repaid is	£2,094.86

Capital outstanding is £10,000 less £2,094.86 = £7,905.14 or calculate the value at 6% of the right to receive 16 more payments of £782.30.

	£782.30
PV £1 pa (YP) for 16 years at 6%	10.1059
	£7,905.85

Capital outstanding is £7,905.85
Therefore, interest will be £7,905.85 × 0.06 = £474.35 and capital repaid in year 10 is £782.30 – £474.35 = £307.95.
In passing it may be observed that the capital repaid in year 1 plus compound interest at 6% will amount to £308 after 9 years:

Year 1 capital	£182.30
Amount of £1 for 9 years at 6%	1.6895
Capital repaid in tenth year	£308 (small error due to rounding)

As interest on a mortgage is based on capital outstanding from year to year, the change in capital repaid from year to year must be at 6%.

	£182.30
Amount of £1 for 24 years at 6%	4.0489
Capital repaid in last year	£738.11

As the annual payment is £782.30 so £782.30 – £738.11 represents the interest due in the last year of the mortgage, namely £44.19.

In many cases the rate of interest will change during the mortgage term. If it does, the mortgagor will usually have the choice of varying the term or repaying at an adjusted rate.

Example 1.24

Given the figures in example 1.23 advise the borrower on the alternatives available if the interest rate goes down to 5% at the beginning of year 10.
  Either: (a) continue to pay £782.30 a year and repay the mortgage earlier or (b) reduce the annual payment.

(a) *No change in annual payment*
  Amount of capital outstanding at beginning of year 10 (ie the end of year 9) is (as before) £7,905.85.This will be repaid at an annual rate of £782.30.

  £7,905.85 divided by £782.30 = PV £1 pa for *n* years at 5% = 10.11

  Interpolation in the PV £1 pa (YP) tables at 5% gives *n* as between 14 and 15 years, ie the loan is repaid before the due date.

(b) *Change in annual payment*

Capital outstanding is	£7,905.85
Annuity £1 wp for 16 years at 5%	0.09227
	£729.47

A change in interest at the beginning of year 10 means that, including the payment in year 10, there are 16 more payments due. The capital outstanding multiplied by the annuity £1 will purchase for 16 years, or divided by PV £1 pa for 16 years, at 5% (the new rate of interest) will give the new annual payment. This could be called the annual equivalent of £7,905.85 over 16 years at 5%.

Summary

This chapter has explored the six single rate investment functions which form the most common basis of property valuations by the income approach and has indicated how the functions are used to assess both returns on various types of investments and borrowing by way of mortgages. The six basic functions and their associated formulae are set out below :

The amount of £1:
used to calculate compound interest

$$A = (1 + i)^n$$

The amount of £1 pa:
used to calculate the future worth of regular periodic investment

$$A £1 \text{ pa} = \frac{(1 + i)^n - 1}{i}$$

Annual sinking fund:
used to calculate the sum that must be
invested annually to cover a known expense
in the future at a known date

$$\text{ASF} \quad = \quad \frac{i}{(1+i)^n - 1} \quad = \quad \frac{i}{A-1}$$

Present value of £1:

used to calculate the present value of
sums to be received in the future

$$\text{PV £1} \quad = \quad \frac{1}{A} \quad = \quad \frac{1}{(1+i)^n}$$

Present value of £1 PA:

$$\text{PV £1 pa} \quad = \quad \frac{1-\text{PV}}{i} \quad = \quad \frac{1 - \dfrac{1}{(1+i)^n}}{i}$$

Also known as the YP and used to
calculate the present value of a
series of payments

$$\text{or} \qquad \frac{1}{i+\text{SF}}$$

Annuity £1 will purchase:
used to calculate the annual sum
given in exchange for an initial
amount of capital

$$\text{Annuity £1 wp} = i + \text{SF}$$

Spreadsheet User 1

Spreadsheets are particularly useful for exploring the valuation tables and investment concepts outlined in this and the other chapters in Part One.

These pages are designed for current but introductory users of Microsoft Excel or Lotus 1-2-3 to help them use the spreadsheet as both a valuation tool and a means of exploring and understanding the concepts of valuation more fully.

Project 1

Excel or Lotus can be used very effectively to generate your own set of valuation tables, which you can personalise to your own requirements, layout and style.

The example below shows how to construct a table for the Amount of £1 function. Only a small section of the spreadsheet is shown in the example: you should create the table for values of i between say 2% and 25% at 0.5% intervals and for 100 years. This project demonstrates the power of spreadsheets to repeat a calculation from a simple copy command. If you generate the whole table as indicated above the Amount of £1 calculation will be performed 4,700 times!

Spreadsheet construction tips

- Plan the table carefully first, decide on its layout and presentation.
- Consider the logical sequence of the calculation and the parentheses required in the formula.

- You must fully understand the nature of relative and absolute cell references in order to understand how the single copy command generates the table by looking up the appropriate values of *i* and *n* in column A and row 2.
- The symbol ^ is used to indicate to the power of.
- The table will require substantial compression to print on a single page. This is simple in Excel by clicking on the option to fit a single page in the Print dialogue box.
- In the example below, the value of *i* is divided by 100 to express the value as a percentage as required in all valuation formulae. You could alternatively format the cells in Row 2 as a percentage and enter the values as 0.02, 0.025, etc. The cell will be displayed as 2%, 2.5%, etc and you can remove the /100 from the formula.

The following table shows the start of the Amount of £1 table. The whole table can be generated by copying the formula in cell C2 as a block for the whole table starting at cell C2 and extending for your chosen range of values of *i* and *n*.

Note: You only have to enter the formula in cell C4 and copy to it both across and down to generate your table The formulae contained in the other cells are shown only to illustrate how the formula changes when it is copied.

	A	B	C	D	E
1		**Interest (%)**			
2		**2**	**2.5**	**3**	**3.5**
3	Years				
4	1	**(1+(C$2/100))^$A4**	(1+(D$2/100))^$A4	(1+(E$2/100))^$A4	(1+(F$2/100))^$A4
5	2	(1+(C$2/100))^$A5	(1+(D$2/100))^$A5	(1+(E$2/100))^$A5	(1+(F$2/100))^$A5
6	3	(1+(C$2/100))^$A6	(1+(D$2/100))^$A6	(1+(E$2/100))^$A6	(1+(F$2/100))^$A6
7	4	(1+(C$2/100))^$A7	(1+(D$2/100))^$A7	(1+(E$2/100))^$A7	(1+(F$2/100))^$A7
8	5	(1+(C$2/100))^$A8	(1+(D$2/100))^$A8	(1+(E$2/100))^$A8	(1+(F$2/100))^$A8

The start of your Amount of £1 table should look like this (see over).

Arial — 10 — **B** *I* <u>U</u> ≡ ≡ ≡ $ % , .00 .00 ≡ ≡ — ⌄ A ⌄

C4 ▼ *fx* =(1+(C$2/100))^$A4

	A	B	C	D	E	F	G	H	I	J
1			Interest (%)							
2	Years		2	2.5	3	3.5	4	4.5	5	5.5
3										
4	1		1.02	1.025	1.03	1.035	1.04	1.045	1.05	1.055
5	2		1.0404	1.050625	1.0609	1.071225	1.0816	1.092025	1.1025	1.113025
6	3		1.061208	1.076891	1.092727	1.108718	1.124864	1.141166	1.157625	1.174241
7	4		1.08243216	1.103813	1.125509	1.147523	1.169859	1.192519	1.215506	1.238825
8	5		1.104080803	1.131408	1.159274	1.187686	1.216653	1.246182	1.276282	1.30696
9	6		1.126162419	1.159693	1.194052	1.229255	1.265319	1.30226	1.340096	1.378843
10	7		1.148685668	1.188686	1.229874	1.272279	1.315932	1.360862	1.4071	1.454679
11	8		1.171659381	1.218403	1.26677	1.316809	1.368569	1.422101	1.477455	1.534687
12	9		1.195092569	1.248863	1.304773	1.362897	1.423312	1.486095	1.551328	1.619094
13	10		1.21899442	1.280085	1.343916	1.410599	1.480244	1.552969	1.628895	1.708144
14										

Dual Rate Tables

As noted in the preface dual rate methodology is no longer considered to be appropriate for the valuation of any interest in property. This chapter is retained for completeness, for the benefit of educators who find the subject-matter to be of some 'academic' interest and for the benefit of those valuers who still use the method for the valuation of leasehold interests in property (see chapter 7).

In single rate annuity and years' purchase tables the sinking fund inherent in the formula is providing for replacement *of* capital at the same rate per cent as the remunerative or investor's rate of return *on* capital. As such the sinking fund is notional and ensures that the correct value is assigned to the investment. Single rate is based on the principle of the rate of return being the internal rate of return (IRR). This provides for the return on money invested to be considered as a return on the amount of capital owed, or outstanding, from year to year. This principle has already been illustrated in relation to annuities and mortgages. Dual rate is based on the principle of the rate of return remaining constant from year to year while the recovery, or redemption of capital occurs through the regular reinvestment of a part of the income in a separate sinking fund which may be accumulating at the same rate as the rate of return on capital or, more typically, at a lower insurable and totally safe rate of interest.

Despite the fact that investors rarely if ever provide for replacement of capital in this way a method developed from the start of the 20th century for the valuation of limited term property investments based upon this sinking fund concept.

The technique is only used by some valuers to value wasting assets such as leasehold interests, it is not the basis of assessing value or price of other time limited investments such as dated government stocks and the application of the method with a specific low sinking fund rate has been, for the most part, restricted in use to the UK. A freehold interest in property could be likened to saving in a building society (the principal is retained in the ownership of the investor) and all income represents a return on capital employed. A leasehold interest is a terminating or wasting asset and comes to an end after a given number of years. The sum originally invested is spent in return for an income for a given number of years. At the end of that time nothing remains. It is just like a mortgage.

The valuer is normally under instruction to assess the market value of an interest in property. A principle of market valuation is comparison and the responsibility of the valuer is to analyse the market so as to be able to assess the market value of other property by comparison. The only tool of analysis and comparison in the investment market is the internal rate of return (single rate) as there can only ever be one IRR for an investment under normal circumstances. To analyse the sale price of a time limited interest in property on a dual rate basis requires the analyst to assume the sinking fund rate (known as the accumulative rate). That assumption then determines the rate of return or remunerative rate. It is thus possible for different analysts to arrive at different views as to the return from a property creating problems of market comparison. The valuer has a further responsibility which is to ensure that the methods of market valuation mirror the behaviour of the buyers and sellers in a particular market. As there is no evidence of investors in property reinvesting in sinking funds there would appear to be no justification for valuers to make such assumptions in their valuation of property.

These arguments are ignored by those valuers whose education and training has taught them that dual rate must be used. The rest of this chapter restates the counter arguments put forward for the use of dual rate and explains how dual rate operates. Readers are referred to A Baum and N Crosby (1995) *Property Investment Appraisal*, for a complete treatise on dual rate methodology.

The dual rate

In order to compare the return from an investment in a leasehold interest with an investment in a freehold interest it is argued that the original outlay must be returned at the end of the lease so that a similar and equal income flow may be acquired. This process may continue into perpetuity so that a perpetual income may be enjoyed and the return becomes comparable with that receivable from a freehold investment.

There would seem to be no problem at first glance because a single rate YP includes a sinking fund to replace initial outlay. But two problems are encountered if an actual sinking fund is to be arranged. While a single rate YP assumes that the sinking fund accumulates at the same rate as the yield from the investment, the rate of interest at which a sinking fund will accumulate does not necessarily relate to the yield given by the investment itself. Two rates may thus be needed. Second, the sinking fund must replace the initial capital outlay and so the possible effect of tax on that part of the income that represents capital replacement cannot be ignored; nor can the effect of tax on sinking fund accumulations.

Because the accumulated sinking fund must be available at the required point in time with no doubt as to its security, a safe rate of interest, often below 4% net of tax, has always been assumed. The overriding question lies not with the rate per cent but with the concept.

As already stated, there is no reason to suppose that the 'accumulative rate' will equate with the yield from the investment or the 'remunerative rate'. Where it does not, the YP figure will be 'dual rate'.

The only formula for a dual rate YP catering for the difference in accumulative and remunerative rates is:

$$\frac{1}{i + sf}$$

The formula

$$\frac{1 - PV}{i}$$

can only be used in the case of a single rate YP;

$$\frac{1}{i + sf}$$

can be used for single or dual rates.

Example 2.1

Value a limited income of £1500 pa for 6 years where your client requires a yield of 11% and the best safe accumulative rate is 3%. Show how a return on and a return of capital are received.

Income	£1,500.00
YP for 6 years at 11% and 3%	3.7793
Capital Value	£5,668.95

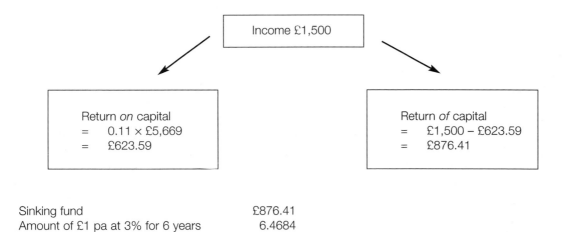

Sinking fund	£876.41
Amount of £1 pa at 3% for 6 years	6.4684
CV	£5,668.97

Dual rate YP and tax

The use of dual rate YP assumes that sinking funds are taken out in practice. Such sinking funds are designed to replace the initial outlay on an investment. Because the sinking fund has to perform this function without question it is assumed to accumulate at a net-of-tax rate.

If an investor pays tax at a rate of say 25%, and the rate of interest earned by the sinking fund is subject to this rate of tax, then a gross accumulative rate will be reduced to a net accumulative rate.

For example, where an ASF is £150 pa and the gross accumulative rate is 5% with tax at 25%, then:

After one year the interest earned is:
£150 × 5% = £7.50 but tax at 25% reduces this to (0.75 × £7.50) = £5.625.
£5.625 is only 3.75% of £150 so the gross rate of 5% has been reduced to a net accumulative rate of 3.75%.

Such a calculation is easily accomplished by applying a tax adjustment factor of $(1 - t)$ to the original gross rate (where t = the rate of tax expressed as a decimal).

If the required sinking fund instalment is calculated on gross instead of net rates of interest it will simply be inadequate whenever interest on the sinking fund is taxed.

Example 2.2

£1,000 must be replaced within 10 years. The accumulative rate is 6%, calculate the Annual Sinking Fund.

Sum required	£1,000
ASF to replace £1 in 10 years at 6%	0.075868
	£75.87

However the sinking fund is taxed at 25%. It will therefore actually accumulate at 6% multiplied by the tax adjustment factor of $(1 - t)$:

=	6% $(1 - t)$
=	6 (0.75)
=	4.5%

ASF	£75.87
A £1 pa for 10 years at 4.5%	12.29
Capital replaced	£932.44

The sinking fund is insufficient to replace the initial outlay of £1,000, due to the effect of tax on the sinking fund accumulation. The SF must be calculated in the light of the tax rate:

Sum required	£1,000
ASF to replace £1 in 10 years at 4.5%	0.08137
ASF	£81.37
A £1 pa for 10 years at 4.5%	12.29
Capital replaced	£1,000

The sinking fund has this time been correctly calculated to accumulate after the effect of 25% tax on the interest accumulating in the sinking fund. Accumulative rates must be net of tax to compensate for this first effect that tax has on the accumulation of the sinking fund.

Tax also affects income from property. Rates of return from most investments are quoted gross of tax, because individual tax rates vary and net-of-tax comparisons may, as a result, be meaningless. The remunerative rate i% in a dual rate YP is therefore a gross rate of interest.

But a sinking fund is tied to the principle that it must actually replace the initial capital outlay so that a comparable investment may be purchased. The effect of tax on the income cannot therefore be ignored.

Example 2.3a

Value a profit rent of £2,000 pa (see Chapter 7) receivable for 10 years using a remunerative rate of 10% gross and an accumulative rate of 3% net. The investor pays tax at 25p in the £ on all property income. Show how the calculation is affected.

Ignoring tax on income	£2,000
YP for 10 years at 10% and 3%	5.341
Capital value	£10,682

But income is taxed at 25p in the £1:

Net income		
	=	£2,000 × (1 − t)
	=	£2,000 (1 − 0.25)
	=	£2,000 (0.75)
	=	£1,500

From this net income a net remunerative rate of 7.5% [10% (1 − t)] and a sinking fund to replace the initial capital outlay must be found.

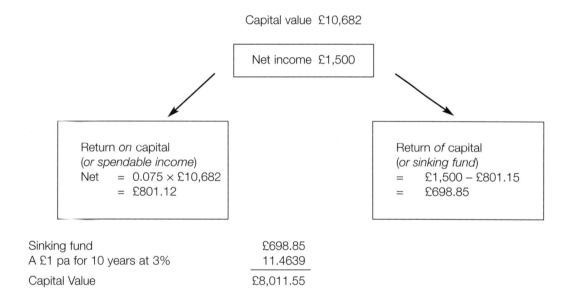

Capital value £10,682

Net income £1,500

Return *on* capital (or spendable income)	Return *of* capital (or sinking fund)
Net = 0.075 × £10,682	= £1,500 − £801.15
= £801.12	= £698.85

Sinking fund	£698.85
A £1 pa for 10 years at 3%	11.4639
Capital Value	£8,011.55

The sum calculated above fails to replace the initial outlay of £10,682.

Why?

Answer: As the income is reduced by 25%, both the spendable income and the sinking fund must be reduced by 25%. The spendable income then becomes a net spendable income representing a net return on capital and still conforms to the investor's requirements. But the net sinking fund must replace £10,682, the fact that the gross sinking fund would notionally replace the initial outlay is no comfort for the investor left several thousand pounds short.

It must therefore be ensured that the net sinking fund still replaces the initial outlay.

Thus:

Gross sinking fund × (1 − *t*)	= net sinking fund

$$\text{Gross sinking fund} = netSF \times \frac{1}{(1 - t)}$$

If, having calculated the desired amount that should remain as a net sinking fund, this amount is multiplied by the 'grossing-up' factor of $1/(1 - t)$ then the required amount of net sinking fund will remain available after tax.

Clearly therefore more income must be set aside as gross sinking fund. This means that the amount of income remaining as spendable income will be reduced, and the investor's requirements of a 10% return will not be fulfilled. A new valuation is therefore required, reducing the price paid so that a grossed-up sinking fund may be provided and a 10% return (gross) can still be attained.

A new YP figure must be calculated using a 10% remunerative rate, a 3% net accumulative rate, and a grossing-up factor applied to the SF of $1/(1 - t)$.

The dual rate YP formula adjusted for tax is therefore
$$\frac{1}{i + \left(SF \times \dfrac{1}{(1 - t)}\right)}$$

In this case this becomes

$$\frac{1}{0.10 + \left[0.08723 \times \dfrac{1}{1 - 0.25} \right]}$$

$$= \frac{1}{0.10 + \left[0.08723 \times \dfrac{1}{0.75} \right]}$$

$$= \frac{1}{0.10 + 0.11630}$$

$$= \frac{1}{0.2163}$$

$$= 4.6232$$

Example 2.3 b

Revaluation of Example 2.3a using the dual rate YP formula :

Income	£2,000
YP for 10 years at 10% and 3% adj tax at 25%	4.6232
Capital value	£9,246

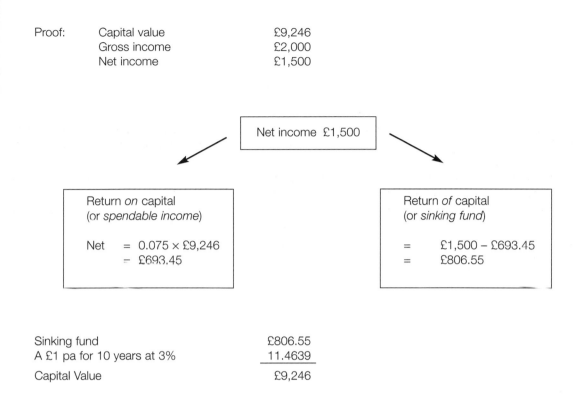

Proof: Capital value   £9,246
              Gross income   £2,000
              Net income   £1,500

Net income £1,500

Return *on* capital
(or *spendable income*)

Net = 0.075 × £9,246
      − £693.45

Return *of* capital
(or *sinking fund*)

= £1,500 − £693.45
= £806.55

Sinking fund	£806.55
A £1 pa for 10 years at 3%	11.4639
Capital Value	£9,246

The result of using a tax-adjusted YP has been to reduce the capital value from £10,682 to £9,246. This has enabled the investor to gross-up the sinking fund to compensate for the effect of income tax and leave enough income after tax to provide a 10% gross and a 7.5% *net* return on capital.

On a single rate basis a sale of an investment income of £2,000 a year for £9,246 is simply analysed to assess the internal rate of return (see Chapter 4). In this instance the IRR is 17.21% or on a net of tax basis with tax at 25%, which reduces the £2,000 to £1,500, the IRR is 9.93%. In practice as valuation is an imprecise science valuers would tend to round the comparable information to 10% net. This single rate basis translated to normal investment terminology of internal rates of return instantly communicates with investors. Terms such as dual rate adjusted for tax have no meaning to investors.

Summary

This chapter has explored the need to consider the replacement of capital where an investment has a limited life, such as a leasehold property investment where the value reduces as it approaches termination. The chapter illustrates the theory of dual rate valuations which provide the mechanism for replacement of capital in order that a further investment can be purchased when the existing investment terminates. In addition the effect of taxation on property incomes has been explored, showing the need to consider the effect of tax on both the spendable income and the sinking fund.

The formula for a dual rate YP adjusted for tax is:

$$\frac{1}{i + \left(SF \times \dfrac{1}{(1 - t)} \right)}$$

The effect of tax on a sinking fund is twofold.

1. *Tax is levied on the interest accumulating on the sinking fund.* To allow for this a *net accumulative rate* must be used.
2. *Tax is levied on the income from which a sinking fund is drawn* So that the correct net sinking fund remains after tax a grossing-up factor is used. A tax-adjusted dual rate YP caters for this, and ensures that the correct valuation is made.
3. The use of dual rate adjusted for tax has been criticised as an inappropriate method of market analysis and market valuation as it does not mirror investor behaviour, can not be readily derived from market sales unless assumptions as to the accumulative rate are made, and does not communicate meaningful information to investors.

Spreadsheet User 2

In the last chapter spreadsheet users were invited to generate their own set of valuation tables using the powerful copy function to generate a whole table from a single formula. However, spreadsheets can be used in the way set out in the project below to make tables redundant by incorporating the valuation mathematics to generate the capital value for any given income, interest rate, sinking fund rate, period and rate of tax.

Project 1

Using Lotus or Excel create an area for entry of the variables and display of the answers.
An example is indicated below.

	A	B	C	D	E
1	Valuation Spreadsheet to calculate the capital value of a terminable income which is taxable :				
2					
3					
4	ENTER THE FOLLOWING VALUATION DATA:				
5					
6				Income £	
7				Period (n) (years)	
8				SF Rate (i_s) %	
9				Interest rate (i) %	
10				Tax rate (t) %	
11					
12				CAPITAL VALUE:	

In cell F12 you will need the formula for dual rate YP adjusted for tax which can be expressed with reference to the above cell layout as :

E6*(1/(E9 + (((E8 / (((1 + E8)^E7) – 1)) * ((1 / (1 – E10))))))

Note : The above assumes that the cells E8,E9 & E10 are in a percentage format (see spreadsheet user 1); if not the values in the above formula require division by 100.

Your Capital Value calculator spreadsheet should look like this.

Microsoft Excel - Book1				

File   Edit   View   Insert   Format   Tools   Data   Window   Help

Arial   ▾ 10 ▾   **B** *I* U   ≡ ≡ ≡   % ,

G16   ▾   *fx*

	A	B	C	D	E
1	Valuation Spreadsheet to calculate the capital value of a terminable				
2	income which is taxable				
3					
4	ENTER THE FOLOWING VALUATION DATA :				
5					
6				Income £	2000
7				Period (*n*) (years)	10
8				SF Rate (*i*s) %	3.000%
9				Interest Rate (*i*) %	10.000%
10				Tax Rate (*t*) %	25.000%
11					
12				CAPITAL VALUE	9246.1032
13					

More Advanced Concepts

The interrelationship of tables

A valuation student's knowledge of the interrelationship of the tables is sometimes tested by problems requiring the use of information provided by one particular table to calculate a related figure from another table. A knowledge of the formulae is essential. It can be seen that the amount of £1 $[(1+i)^n$ or $(A)]$ is present in each formula:

1 PV $$\frac{1}{A}$$

2 A £1 pa $$\frac{A-1}{i}$$

3 ASF $$\frac{i}{A-1}$$

4 YP $$\frac{1-\dfrac{1}{A}}{i} \qquad \text{or} \qquad \frac{1}{i+\dfrac{i}{A-1}}$$

5 A £1 wp $$i+\frac{i}{A-1}$$

6 A $$(1+i)^n$$

It can also be seen from the above that the six tables fall into three sets of reciprocals: 1 and 6; 2 and 3; 4 and 5.

It is through the understanding of this and the nature of the formula that the key to the solution of the following type of problem is found.

Example 3.1

Showing your workings, calculate to four places of decimals:

(i) PV £1 pa at 5% for 20 years given the PV £1 for 20 years at 5% is 0.37689.

$$PV = \frac{1}{A} \text{ and } PV\pounds1 \text{ pa} = \frac{1 - PV}{i}$$

$$PV \pounds1 \text{ pa for 20 years at 5\%} = \frac{1 - 0.37689}{0.05} = \frac{0.62311}{0.05} = 12.4622$$

(ii) ASF necessary to produce £1 after 15 years at 3% given that the Amount of £1 for 15 years at 3% is 1.558.

$$ASF = \frac{i}{A - 1} = \frac{0.03}{1.558 - 1} = 0.0538$$

(iii) YP 70 years at 5% and 2.5% given that the ASF to produce £1 in 70 years at 2.5% is 0.0053971.

$$YP = \frac{i}{i + ASF} = \frac{i}{0.05 + 0.0053971} = \frac{i}{0.0553971} = 18.0515$$

Note *i* is the remunerative rate and ASF (SF) is the annual sinking fund to produce £1 for n years — in this case 70 years — at the accumulative rate *i*% which in this case is 2.5%.

Some problems may require more than simple substitution.

Example 3.2

Given that the A £1 pa for 7 years at 4% is 7.8983, find the YP for 7 years at 4%.

$$YP = \frac{1}{i + ASF}$$

$$ASF = \frac{i}{A - 1} \text{ and } A\pounds1 \text{ pa} = \frac{A - 1}{i} \text{ and therefore } A\pounds1 \text{ pa} = \frac{1}{ASF}$$

So the value of ASF in the YP formula is the reciprocal of·the Amount of £1 pa for 7 years at 4% which is given as 7.8983.

Therefore

$$YP = \cfrac{1}{0.04 + \cfrac{1}{7.8983}} = \frac{1}{0.04 + 0.1266} = \frac{1}{0.1666} = 6.002$$

The problem may be complicated further by changing the relevant number of years. This obviously presents a considerable difficulty. It is no longer sufficient just to isolate A, as it will represent the wrong number of years in the annuity figure.

In the compound interest formula $(1 + i)^n$, n is the exponent and the expression means that $(1 + i)$ is multiplied by itself n times. Exponents have certain properties and a knowledge of these can be useful when manipulating the various valuation formulae.

The properties of exponents

1   Any number raised to the zero power equals 1.

2   A fractional exponent is the root of a number:

$(1 + i)^{1/2}$ $=$ $(1 + i)^{0.5}$ $=$ $^2\sqrt{(1 + i)}$

similarly

$(1 + i)^{1/5}$ $=$ $(1 + i)^{0.2}$ $=$ $^5\sqrt{(1 + i)}$

3   When a number being raised to a power is multiplied by itself the exponents are added:

$(1 + i)^4 \times (1 + i)^6$ $=$ $(1 + i)^{10}$ and if being divided by itself they are subtracted
$(1 + i)^{10} \div (1 + i)^5$ $=$ $(1 + i)^5$

Example 3.3

Given that ASF 16 years at 4% is 0.04582 find the Annuity £1 will purchase for 14 years at 4%.

$$\frac{A£1 \text{ for 16 years at 4\%}}{A£1 \text{ for 2 years at 4\%}} = A£1 \text{ for 14 years at 4\%}$$

$$ASF = \frac{i}{A - 1} = 0.04582$$

$$A - 1 = \frac{0.04}{0.04582} = 0.873$$

$$A = 0.0873 + 1 = 1.873$$

$$\frac{1.873}{(1 + i)^2} = A£1 \text{ for 14 years at 4\%}$$

$$\frac{1.873}{(1.04)^2} = \frac{1.873}{1.0816} = 1.7317$$

Thus A £1 wp for 14 years at 4% $= i + \dfrac{i}{A - 1}$ and now substituting from above the solution becomes:

$$= 0.04 + \frac{0.04}{1.7317 - 1} = 0.04 + \frac{0.04}{0.7317} = 0.04 + 0.05467 = 0.09467$$

Some valuation tutors still insist on testing this competence even further by using false figures in the set problems thereby, in their view, fully testing their students' mathematical skills. The advent of financial calculators, Excel and industrial valuation software packages while not diminishing the need to understand the construction of the six financial functions of £1 have largely replaced the need for this level of appreciation of the interrelationships with the opportunity to undertake more useful accurate application of the functions to situations where compounding or discounting is more properly required at intervals other than yearly. This step in the use of the six functions requires an awareness of the differences between nominal and effective rates of interest.

Nominal and effective rates of interest

The rates of return provided by different forms of saving and from different forms of investment are usually compared by means of the annual rate of interest being paid or earned. For example banks and building societies usually quote the annual rate of interest that will be earned in savings accounts, such as 10% pa, representing the amount of interest that will be added to each £100 of savings left in the account for a full 12 months. Sometimes, however, it is necessary to check whether the rate of interest quoted is nominal or effective. The distinction is about knowing the frequency with which interest is added to a savings account and how it is added.

For example if a building society pays interest at 10% a year and it is added at the end of each complete year with the affect that £10 is added to the account for every £100 held for a year in that account then the rate of 10% is both the nominal rate of interest and the effective rate of interest.

Consider, on the other hand, building society B which pays 10% pa with interest paid half-yearly. 10% per annum paid half-yearly at a rate of 5% every six months means a higher effective rate than the nominal 10% because the interest paid after six months will itself earn interest over the second half of the year. The total accumulation of £1 invested in an account for one year will therefore be £$(1.05)^2$ = £1.1025. Interest is 10.25p, accumulated on £1 invested. The effective rate of interest is therefore 10.25%.

However if the 10% is quoted as the AER or as an effective rate of interest but is actually credited every six months then it is important to recognise that 5% or £5 for every £100 will not be added after the first six months. Instead the savings provider will calculate the effective rate for each six months. So the amount earned every six months can be found by using the amount of £1 formula. The effective 10% becomes 4.88% payable every six months because:

$$(1 + i)^n = (1 + i)^2 = 1.10$$

The interest is to be paid twice a year but after the interest is added and after interest is added to the interest earned in the first six months, ie after compounding has occurred, the total per £100 is not to exceed 10% or 0.10 per £1. The problem is to calculate the six monthly rate of interest.

$(1 + i)^2 = 1.10 \therefore (1 + i) = \sqrt[2]{1.10} = 1.0488$

$i = 1.0488 - 1 = 0.0488$

$0.0488 \times 100 = 4.88\%$

If interest is to be paid quarterly the rate of interest per quarter becomes 2.411%. This can be checked by substituting in the Amount of £1 formula:

$(1 + i)^4 = (1 + 0.02411)^4 = (1.02411)^4 = 1.10$

Investments are therefore best compared by means of the annual equivalent rate of interest. Building societies, banks and all other finance companies are required to disclose this annual equivalent rate (AER). The AER may be based on sums which include arrangement costs.

This principle leads to the conclusion that given a specific effective rate of interest per year the effective rate per period will get ever smaller as the interest earning period gets smaller. For many purposes a daily rate is used and interest is earned and credited on a daily basis. Similarly certain loans and mortgages will have interest charged on the amount of the loan or debt outstanding from day to day or from month to month. This concept leads naturally to the possibility of continuous compounding.

Continuous compounding

When interest is added more frequently than annually, the compound interest formula $(1 + i)^n$ is adjusted to:

$$\left(1 + \frac{i}{m} \right)^{mn}$$

Where m is the number of times per year that interest is added, i is the nominal rate of interest per year and n is the number of years. Here the more frequently interest is added the greater will be the annual effective rate of interest.

If for example the nominal rate of interest is 12% per year and interest is added monthly then:

$$\left(1 + \frac{0.12}{12} \right)^{12} = (0.01)^{12} = 1.1268$$

The effective rate is 12.68%.

The greater the number of times in the year that interest is added, the greater will be the total sum at the end of the year, but there must be a limit, because while the number of periods becomes infinitely large the rate of interest per period becomes infinitely small.

Potentially, m might tend towards infinity. This would imply the immediate reinvestment of earned interest, or *continuous compounding*.

Mathematically the maximum sum to which £1 could compound in 1 year at 100% pa is given by the following series:

$$1 + \frac{1}{1} + \frac{1}{1 \times 2} + \frac{1}{1 \times 2 \times 3} + \frac{1}{1 \times 2 \times 3 \times 4} \; ... \; \frac{1}{m}$$

which is a convergent series summating to 6 places, to 2.718282 or *e*.

The mathematical proof behind continuous compounding can be found elsewhere, for our purposes it can be accepted that the maximum sum to which £1 can accumulate over *n* years with interest at *i* and with continuous compounding becomes e^{in}.

Example 3.5

To what sum will £1 compound over 2 years at 10% per year nominal rate of interest assuming continuous compounding?

$$e^{in} \;=\; 2.71828^{(0.1)(2)}$$
$$e^{in} \;=\; 2.71828^{(0.2)}$$
$$e^{in} \;=\; 1.211$$

£1 accumulating at 10% per year with interest added annually would only compound to $(1.10)^2$ or £1.21.

$$\text{The present value of £1} \;=\; \frac{1}{(1 + i)^n} \;=\; (1 + i)^{-n}$$

It follows then that where interest is compounding continuously: The present value of £1 $= e^{-in}$

The foregoing demonstrates that the rate of interest used in calculations of compounding and discounting must be the effective rate for the period.

Incomes in advance and non-annual incomes

The issue here is how should a valuer deal with the valuation of incomes received in advance? Many years' purchase valuation tables assume that the unit of income is received at the end of the year but *Parry's* and others now contain Quarterly in advance tables. The same may apply to PV and A £1 pa tables. The Investment Property Forum have made strong pleas for the property market to adopt in advance practices but while investment advice is frequently provided on the basis of the 'true equivalent yield' which reflects the precise timing of rental payments, most valuers undertake, and most valuations are undertaken, on the basis of in arrears assumptions.

However, this is not always a realistic assumption. Rent from property is usually paid in advance. Tables giving 'in arrears' figures may be used alongside common sense to provide related 'in advance' figures.

1   PV £1: normally assumes that the sum is to be received at the end of the year n. If the sum is received at the start of year n instead this will coincide with the end of year (*n* – 1).

Thus the PV of £1 receivable in advance $\quad = \dfrac{1}{(1 + i)^{n-1}}$

2. Amount of £1 pa normally assumes that each £1 is invested at the end of each year. If this becomes the start of each year instead an extra £1 will accumulate for n years — but the £1 paid at the end of the nth year will now be paid at the start of year n. The series is now:

$$(1 + i)^n \; + \; (1 + i)^{n-1} \; ... \; (1 + i)^2 \; + \; (1 + i)$$

which can be summated to:

$$\left[\dfrac{(1 + i)^{n+1} \; - \; 1}{i}\right] \; -1$$

3. Year's purchase or PV £1 pa usually assumes income to be received at the end of each year. But if it comes in advance, the series will read:

$$1 + \dfrac{1}{(1 + i)} + \dfrac{1}{(1 + i)^2} \; ... \; \dfrac{1}{(1 + i)^{n-1}} = \dfrac{1 - \dfrac{1}{(1 + i)^{n-1}} + 1}{i}$$

Example 3.4

Calculate YP for 6 years at 10% in advance.

$$\dfrac{1 - \dfrac{1}{(1 + 0.10)^{6-1}}}{i} + 1 = \dfrac{1 - \dfrac{1}{1.6105}}{0.10} + 1 = 3.79 + 1 = 4.79$$

Alternatively, this could be given by YP for 5 years at 10% + 1: (3.79) + 1 = 4.79, or by YP for 6 years at 10% x (1 + i).

In this book, so far, i has always represented an annual interest rate. But the formulae can be used for alternative interest rate periods which are commonly used in property investment transactions. It must be ensured however that i and n relate to the same time period.

The time period can be anything; so for example if the interest rate i is quarterly, the time period n must represent quarterly periods: ie for 1 year n would equal 4.

Example 3.5

Calculate the amount of £1 in 7 years 6 months at an interest rate of 3% per half year.

In this example, the interest rate (i) is per half year, so n should represent periods of half a year.

$$i \quad = \quad 0.03 \qquad\qquad n \quad = \quad 15 \text{ periods}$$

$$A \quad = \quad (1 + i)^n$$

$$= \quad (1 + 0.03)^{15}$$

$$= \quad 1.557967$$

The same logic may be applied to all six functions as specified at the beginning of this chapter.

Incomes receivable quarterly in advance

It was noted in chapter one that some valuation tables are based on the assumption that income may be received or invested quarterly.

At the current time it is much more common to find income from property (rent) paid quarterly or monthly in advance. *Bowcock, Rose* and *Parry's Valuation Tables* all contain quarterly tables. By far the most common application of such a basis is to the Years' Purchase tables.

The formula for the Years' Purchase to be applied to an income receivable quarterly in advance, single rate, is:

$$YP \quad = \quad \frac{1 - \left[\dfrac{1}{(1 + i)^n} \right]}{4 \left[1 - \dfrac{1}{\sqrt[4]{(1 + i)}} \right]}$$

where i is the annual effective rate of interest.

For example, the single rate Years' Purchase, quarterly in advance, at an annual effective rate of 10% for 20 years, is 9.038, compared with the equivalent annual in arrears years' purchase of 8.5136, reflecting the advantages of receiving income both earlier and more regularly.

This edition does not recommend the use of dual rate for leasehold valuations but out of interest the formula for the years' purchase to be applied to an income receivable quarterly in advance, dual rate with a tax adjustment, is:

$$YP \quad = \quad \frac{1}{4 \left[1 - \dfrac{1}{\sqrt[4]{(1 + i)}} \right] + \dfrac{4 \left[1 - \dfrac{1}{\sqrt[4]{(1 + s)}} \right]}{[(1 + s)^n - 1] \times [1 - t]}}$$

where i is the annual effective remunerative rate, s is the annual effective accumulative rate, and t is the tax rate.

For example, the dual rate years' purchase quarterly in advance, at an annual effective remunerative rate of 10%, an annual effective accumulative rate of 3%, adjusted for tax at 40%, for 20 years is 6.3555, compared with the annual in arrears equivalent figure of 6.1718.

It is strictly correct that such factors should be applied to incomes which are received quarterly in advance, in order that it is demonstrable that the yield indicated by the valuation is actually provided by the investment. The use of other tables might provide valuations which are acceptable in the market, but it should be noted that such valuations are based upon slightly misleading rates of return. The chapter on freehold valuations explores these applications in respect of typical valuation problems. It should be noted here that in all investment calculations it is imperative that the right rate per cent is used on each occasion. Thus in practice it is important to reflect on the nature of the rate per cent and as to whether when switching from annual in arrears to quarterly in advance one is dealing with the relevant effective rates.

Summary

This chapter has demonstrated the interrelationships between the six functions of £1. It has explained the important concept of effective rate of interest per interest earning period and illustrated how the formulae for the six functions of £1 can be amended to cover payments made in advance and where interest is due quarterly or at other frequencies.

The PV of £1 receivable in advance:

$$\frac{1}{(1 + i)^{n-1}}$$

This is used to calculate present value where the sum is received at the start of the year

The amount of £1 pa receivable in advance:

$$\left[\frac{(1 + i)^{n+1} - 1}{i}\right] - 1$$

this is used to calculate the future worth of regular periodic investments made a the start of each year

Present value of £1 pa or YP receivable in advance:

$$\frac{1 - \dfrac{1}{(1 + i)^{n-1}}}{i} + 1$$

used to calculate the present worth of a series of payments received in advance

YP single rate, income receivable quarterly in advance:

$$\frac{1 - \left[\dfrac{1}{(1 + i)^{n}}\right]}{4\left[1 - \dfrac{1}{\sqrt[4]{(1 + i)}}\right]}$$

Spreadsheet User 3

In the previous chapter spreadsheet users were invited to construct a spreadsheet to value an income using the dual rate YP adjusted for tax for any values of *n*, *i*, is and *t*. The same principles of construction can be applied to the concepts and formulae explored in this chapter. The example below is used to construct spreadsheet to value income receivable quarterly in advance. Spreadsheet users should try to apply the solution to all the formulae explored in this chapter.

Project 1

Using Lotus or Excel create an area for entry of the variables and display of the answers.
An example is indicated below:

	A	B	C	D	E	F
1	Valuation Spreadsheet to calculate the capital value of an income					
2	receivable quarterly in advance:					
3						
4	ENTER THE VALUATION DATA:					
5						
6				Income	£	
7				Period (*n*)	(years)	
8			Annual effective interest rate (i)		%	
9						
10						
11				CAPITAL		
12				VALUE:		

In cell F12 you will need the formula for YP applied to an income receivable quarterly in advance, single rate which can be expressed with reference to the above cell layout as :

F6*((1-(1/((1+F8)^F7)))/(4*(1-(1/((1+F8)^0.25)))))

Note: The above assumes that cell F8 is in a percentage format; if not, the values in the above formula require division by 100

At this stage readers may find it useful to construct their own Excel 'Ready Reckoner' Spreadsheet for valuation purposes incorporating those functions of £1 that they feel would be most useful for their studies of the Income Approach.

Project 2

Construct a 'Ready Reckoner' which displays the various values of £1 using the basic six functions of £1 and those additional functions that are most frequently used.

Create the spreadsheet so that it displays the values of £1 and the values for any actual sums that might be required.

Your 'Ready Reckoner' could look like the one set out below:

	File Edit View Insert Format Tools Data Window Help					
	A	B	C	D	E	F
2						
3	INFORMATION	DATA ENTRY:				
4	YEARS	15				
5	%(enter as say 5%)	10%				
6	AMOUNT	£10,000				
7						
8	INVESTMENT FUNCTIONS		Factor per £	Value		
9	Amount £		4.177248169	£41,772.48		
10	Amount £ pa		31.77248169	£317,724.82		
11	Amount £ pa in advance		34.94972986	£349,497.30		
12	Annual Sinking Fund		0.031473777	£314.74		
13	Present Value of £		0.239392049	£2,393.92		
14	Present Value of £pa(YP)		7.606079506	£76,060.80		
15	Present Value of £ per quarter in advance (YP)		8.075796818	£80,757.97		
16	Annuity £ will purchase		0.131473777	£1,314.74		
17	Mortgage repayment pa		0.131473777	£1,314.74		
18						
19	YP in perpetuity		10	£100,000.00		
20	YP in perpetuity in advance		11	£110,000.00		
21	YP quarterly in advance in perpetuity		10.61755509	£106,175.66		
22	YP perpetuity deferred		2.393920494	£23,939.20		
23	YP quarterly in advance in perp.deferred		2.541758272	£25,417.58		
24	YP Dual rate at 4% unadj		6.669285456	£66,692.85		
25	YP Dual rate at 4% adj.tax at 40%		5.457418081	£54,574.18		
26						
27	*The amount will be the amount saved, borrowed, to be accumulated to in a sinking fund, the annual rent etc.*					
28	*The years will in some instances be the period of deferment eg YP perp deferred.*					
29						
30						

Discounted Cash Flow 4

Discounted Cash Flow (DCF) is an aid to the valuation or analysis of any investment producing a cash flow. In its general form, it has two standard products — NPV and IRR.

Net present value (NPV)

Future net benefits receivable from the investment are discounted at a given *'target rate'*. The sum of the discounted benefits is found and the initial cost of the investment deducted from this sum, to leave what is termed the net present value of the investment, which may be positive or negative. A positive NPV implies that a rate of return greater than the target rate is being yielded by the investment; a negative NPV implies that the yield is at a rate of return lower than the target rate. The target rate is the minimum rate which the investor requires in order to make the investment worth while, taking into account the risk involved and all other relevant factors. It will be governed in particular by one factor: the investor's cost of capital.

Investments may require the initial outlay of a large capital sum, and investors will often be forced to borrow money in order to accumulate that sum. The interest to be paid on that loan will be the investor's cost of capital at that time. It is clear that the rate of return from an investment where the initial capital has been borrowed should be at least equal to the cost of capital, or a loss will result.

An alternative way of looking at this is that the investor will always have alternative opportunities for the investment of his capital. Money may be lent quite easily to earn a rate of interest based on the cost of capital. The return from any investment should therefore compare favourably with the opportunity cost of the funds employed, and this will usually be related to the cost of capital.

For these reasons, the target rate should compare well with the cost of capital. From this basis, a positive or negative NPV will be the result of the analysis and upon this result the investment decision may be made.

Example 4.1

Find the NPV of the following project, using a target rate based on a cost of capital of 13%.

Outlay £10,000

	Income	PV £1 at 13%	Discounted sum £
Returns in year 1	5,000	0.8849	4,425
Returns in year 2	4,000	0.7831	3,133
Returns in year 3	6,000	0.6930	4,158
			£11,716
		Less outlay	£10,000
		NPV	£1,716

The investment yields a return of 13% and, in addition, a positive NPV of £1,716. In the absence of other choices, this investment may be accepted because the return exceeds the target rate.

However, it is more usual for the investment decision to be one of choice.

Example 4.2

An investor has £1,400 to invest and has a choice between investment A and B. The following returns are anticipated:

Income flow		A £	B £
Year	1	600	200
	2	400	400
	3	200	400
	4	400	600
	5	400	600

The investor's target rate for both investments is 10%, the investments are mutually exclusive (only one of the two can be undertaken); which investment should be chosen?

Income flow A		Income	PV£1 at 10%	Discounted sum
Year	1	600	0.9091	545
	2	400	0.8264	330
	3	200	0.7513	150
	4	400	0.6830	273
	5	400	0.6209	248
				£1,546
			Less outlay	£1,400
			NPV	£146

Income flow B	*Income*	*PV£1 at 10%*	*Discounted sum*
Year 1	200	0.9091	182
2	400	0.8264	330
3	400	0.7513	300
4	600	0.6830	409
5	600	0.6209	373
			£1,594
		Less outlay	£1,400
		NPV	£194

From this information investment *B* should be chosen. Each investment gives a return of 10% plus a positive NPV: *B* produces the greater *NPV £194* compared to *A £146*.

If either investment had been considered to be subject to more risk, this factor should have been reflected in the choice of target rate.

In example 4.2 both investments involved the same amount of capital or initial outlay. However, this will not always be the case, and, when outlays on mutually exclusive investments differ, the investment decision will not be so simple. An NPV of £200 from an investment costing £250 is considerably more attractive than a similar NPV produced by a £25,000 outlay.

How can this be reflected in an analysis?

A possible approach is to express the NPV as a percentage of the outlay.

Example 4.3

Which of these mutually exclusive investments should be undertaken when the investor's target rate is 10%?

Investment A		Investment B
Outlay £5,000		Outlay £7,000

Income flow		
Year 1	£3,000	£4,000
Year 2	£2,000	£3,000
Year 3	£1,500	£2,000

Income flow A	£	*PV£1 at 10%*	*Discounted sum*
Year 1	3,000	0.9091	2,727
2	2,000	0.8264	1,653
3	1,500	0.7513	1,127
			£5,507
		Less outlay	£5,000
		NPV	£507

Income flow B		£	PV£1 at 10%	Discounted sum
Year	1	4,000	0.9091	3,636
	2	3,000	0.8264	2,479
	3	2,000	0.7513	1,503
				£7,618
			Less outlay	£7,000
			NPV	£618

At first sight, B might appear to be more profitable, but if the NPV is expressed as a percentage of outlay the picture changes.

$$A = \frac{£507}{£5000} \times 100 = 10.14\%$$

$$B = \frac{£618}{£7000} \times 100 = 8.33\%$$

On this basis *A*, and not *B*, should be chosen.

The NPV method is a satisfactory aid in the great majority of investment problems but suffers from one particular disadvantage. The return provided by an investment is expressed in two parts, a rate of return and a cash sum in addition, which represents an extra return. These two parts are expressed in different units which may make certain investments difficult to compare.

This problem is not present in the following method of expressing the results of a DCF analysis.

The internal rate of return (IRR)

This is the discount rate which equates the discounted flow of future benefits with the initial outlay. It produces an NPV of 0 and may be found by the use of various trial discount rates.

Example 4.4

Find the IRR of the following investment.

Outlay £6,000:	Returns	Year 1	£1,024
		Year 2	£4,000
		Year 3	£3,000

Trying 10%:		£	PV£1 at 10%	Discounted sum
Year	1	1,024	0.9091	931
	2	4,000	0.8264	3,306
	3	3,000	0.7513	2,253
				£6,490
			Less outlay	£6,000
			NPV	£490

At a trial rate of 10%, a positive NPV results. £490 is too high — an NPV of 0 is the desired result. The trial rate must be too low, as the future receipts should be discounted to a greater extent.

Trying 16%:

		£	PV£1 at 16%	Discounted sum
Year	1	1,024	0.8621	931
	2	4,000	0.7432	2,972
	3	3,000	0.6407	1,923
				£5,778
			Less outlay	£6,000
			NPV	−£222

This time the receipts have been discounted too much. A negative NPV is the result, so the trial rate is too high. The IRR must be between 10% and 16%

Trying 14%:

		£	PV£1 at 14%	Discounted sum
Year	1	1,024	0.8772	899
	2	4,000	0.7695	3,076
	3	3,000	0.6750	2,025
				£6,000
			Less outlay	£6,000
			NPV	£0,000

As the NPV is £ 0 in example 4.4, the IRR must be 14%.

Calculation of the IRR by the use of trial rates will be difficult when the IRR does not happen to coincide with a round figure, as in the following illustration.

Outlay £4,925

Trying 11%:

		Cash Flow £	PV£1 at 11%	Discounted sum
Year	1	2,000	0.9009	1,802
	2	2,000	0.8116	1,623
	3	2,000	0.7312	1,462
				£4,887
			Less outlay	£4,925
			NPV	−£38

The trial rate is too high:

Trying 10%:

		£	PV£1 at 10%	Discounted sum
Year	1	2,000	0.9091	1,818
	2	2,000	0.8264	1,652
	3	2,000	0.7513	1,503
				£4,973
			Less outlay	£4,925
			NPV	£48

The IRR is therefore between 10% and 11%.

Published tables may not give PV figures between 10% and 11%; the continued use of trial rates to make a more accurate estimation of the IRR will therefore be impracticable. The analysis has shown that the IRR lies between 10% and 11%, but the accuracy of such an analysis is limited.

A graph of NPVs plotted against trial rates will usually take the form shown in Figure 4.1 below.

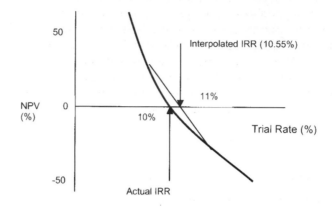

Figure 4.1 NPVs vs trial rates

The graph shows how the IRR (producing an NPV of 0) lies between 10% and 11%.

If the graph is drawn accurately, it will be possible to estimate the IRR in this way. This process is both difficult and time-consuming and will not guarantee complete accuracy. The graph takes the shape of a gentle curve and although a straight line may be assumed between two sufficiently close trial rates and the IRR estimated to a fair degree of accuracy, it is not precise. This process is known as linear interpolation.

Linear interpolation can also be carried out by the use of the following formula, which simply assumes a straight-line relationship between the trial rates and the resulting NPVs.

$$IRR = ltr + \frac{NPVltr}{NPVltr + NPVhtr} \times (htr - ltr)$$

Lower trial rate: *ltr*
Higher trial rate: *htr*

In this case the lower trial rate is 10%; the higher trial rate is 11%; the NPV at the lower trial rate is £48 and the NPV at the higher trial rate is –£38.

The difference in trial rates is 1%.

$$IRR = 10\% + \left(\frac{48}{48 + 38}\right) \times (11\% - 10\%)$$

$$= 10 + \left(\frac{48}{86} \times 1\right)$$

$$= \quad 10 + 0.558$$
$$= \quad 10.558\%$$

This result is a more accurate representation of the geographical interpolation, which yielded a result of 10.55%. This method will provide a satisfactory answer in the majority of cases. However, it must be borne in mind that a graph of NPVs plotted against trial rates will result in a curve, and not a straight line between two points. Because of this, linear interpolation is inaccurate to a certain extent, and the result should not be expressed to too many decimal places.

An IRR of 10.56% will be sufficiently reliable and precise for most purposes. Such calculations, however, are more readily undertaken using an investment calculator, the correct figure being 10.56087452%. The effect of the straight line assumption can be seen in the difference between the figure of 10.56% and the previous figure of 10.558%, figures which are significant in the money markets dealing in billions of pounds.

Comparative use of NPV and IRR

When analysing a range of investments NPV and IRR results may be compared with one another. In some cases conflicting results may arise and additional analysis may be needed to determine the best choice of investment.

Example 4.5

On the basis of NPV and IRR methods rank the following investments, using a rate of 10% in each case.

Investment A		Investment B	
Outlay £5,000		**Outlay £10,000**	
Income:	£	Income:	£
Year 1	600	Year 1	3,342
Year 2	2,000	Year 2	3,342
Year 3	4,000	Year 3	3,342
Year 4	585	Year 4	3,342

(a) NPV analysis using a trial rate 10%

Investment A:		£	PV£1 at 10%	Discounted sum
Year	1	600	0.9091	545
	2	2,000	0.8264	1,653
	3	4,000	0.7513	3,005
	4	585	0.6830	399
				£5,602
			Less outlay	£5,000
			NPV	£602

Investment B:	£	*PV£1 at 10%*	*Discounted sum*
Year 1	3,342	0.9091	3,120
2	3,342	0.8264	2,836
3	3,342	0.7513	2,578
4	3,342	0.6830	2,344
			£10,878
		Less outlay	£10,000
		NPV	£878

On the basis of the NPV analysis, the investor should choose investment B which has the highest NPV.

(b) IRR analysis: It is clear from the NPV analysis that each investment has an IRR exceeding 10% because the NPV is positive.

Investment A:

Trying 15%	£	*PV£1 at 15%*	*Discounted sum*
Year 1	600	0.8696	522
2	2,000	0.7561	1,512
3	4,000	0.6575	2,632
4	585	0.5718	334
			£5,000
		Less outlay	£5,000
		NPV	£0,000

Investment B:

Trying 14%	£	*PV£1 at 14%*	*Discounted sum*
Year 1	3,432	0.8772	3,011
2	3,432	0.7695	2,639
3	3,432	0.6750	2,317
4	3,432	0.5921	2,033
			£10,000
		Less outlay	£10,000
		NPV	£0,000

On the basis of the IRR analysis, the investor should choose investment A, because it has the highest IRR.

In this example the IRR and NPV methods give different rankings.

Which investment should be chosen?

In investment B, an extra £5,000 has been employed. This can produce certain extra benefits. To analyse the situation further it is necessary to tabulate the difference in cash flows between investment A and B.

	A	B	B–A
Outlay	£5,000	£10,000	£5,000
Receipts:			
Year 1	£600	£3,432	£2,832
Year 2	£2,000	£3,432	£1,432
Year 3	£4,000	£3,432	–£568
Year 4	£584	£3,432	£2,848

Incremental analysis

The final column could be called the increment of B over A. This in itself becomes a cash flow on which an NPV or IRR analysis could be carried out. The cost of capital is 10%, and, as it has been assumed to represent the investor's target rate, it is also assumed that this is a rate of return that could be earned elsewhere.

If the increment earns a return of less than 10%, the investor would be wise to invest £5,000 in project A and obtain 10% interest elsewhere with the remaining £5,000.

Another way of looking at this is to remember that the cost of capital could also represent the cost of borrowing money. If the investor's £10,000 has been borrowed, a return of less than 10% on the increment of £5,000 of project B over project A means that the loan charges on this £5,000 are not being covered and the loan of the second £5,000 was not worthwhile.

However, if the increment can be shown to produce a return in excess of 10%, the loan charges will be covered. Investment B thus uses the whole £10,000 to good effect. If project A were chosen, the extra £5,000 could only earn 10%, as it is assumed that no return in excess of the cost of capital could be earned without incurring an extra element of risk.

Such analysis is called incremental analysis.

Incremental flow:

		£	PV£1 at 10%	Discounted sum
Year	1	2,832	0.9091	2,575
	2	1,432	0.8264	1,183
	3	- 568	0.7513	–427
	4	2,848	0.6830	1,945
				£5,276
				–£5,000
			NPV[1]	£276

[1] Any project incorporating frequent sign changes can produce more than one IRR due to the polynomial nature of the underlying equation.

The IRR of the increment exceeds 10%, so investment B should be chosen. The first £5,000 employed is as profitable as it would be if used in investment A; the second £5,000 is used more profitably than is possible elsewhere. B is therefore preferable to the investment of £5,000 at the cost of capital rate of interest of 10%.

Discounted cash flow tables in certain sets of valuation tables are simply extensions of the present value of £1 and present value of £1 pa tables, this illustrates the point that DCF is nothing more than a present value exercise.

Summary

This chapter has demonstrated the use of the PV function to assess the acceptability of investment opportunities measured against an investors target rate of return. In those cases where the results of an NPV and an IRR analysis conflict an incremental analysis may be undertaken to assess the return on any additional capital used.

The discounted cash flow process has two key components.

- A Net Present Value calculation (NPV) discounts the net benefits of an investment using the present value function at the investor's target rate or opportunity cost of capital rate.
  The initial outlay is deducted from the NPV and if the resultant sum is positive then the investment is acceptable at the target rate used.
- An Internal Rate of Return (IRR) represents the investments return and is calculated by trial and error or through the use of an iterative process using investment calculators or spreadsheets. Generally the higher the IRR the better the investment opportunity given the same risk assessment.
- Care needs to be taken in accurately assessing the benefits and timing of such benefits, including any terminal or resale value at the end of the cash flow.
- Cash flows with negative sums arising during the life of the investment may produce multiple values of IRR.

Spreadsheet User 4

Spreadsheets are ideally suited to financial modelling, the application of discounted cash flow solutions and the calculation of internal rates of return. Indeed most spreadsheets have financial functions such as NPV and IRR built in to save time and trouble. However, as explained below the use of the built-in functions must be undertaken carefully as they may not operate in the way in which the user expects.

Project 1: Calculating the NPV

You are evaluating a scheme which will cost £80,000 immediately and generates the following cash flow at the end of each year:

Year 1	£10,000
Year 2	£25,000
Year 3	£35,000
Year 4	£30,000
Year 5	£20,000

Calculate the Net Present Value of the Scheme and the Internal Rate of Return.

1   Construct a spreadsheet which calculates the Net Present Value.
    Use the Present Value of £1 formula in cell E5 designed to pick up the discount rate (*i*) from the amount entered by the user in Cell 3 and copy this formula to cells E6:E9, ie IN CELL E5 should be the formula =1/((1+(F3))^C5).
    In cell F5 enter the formula D5*E5 and copy to cells F6:F9.
    To calculate the Net Present Value in cell F10 use the formula = sum(F4:F9)
    (The use of the in-built Net Present Value and Internal Rate of Return functions are explored below).

An example is indicated below:

	A	B	C	D	E	E	F
1							
2			YEAR			PV £1 at	Discounted
3					Discount Rate =	10%	Sum
4		Expenditure		–80000			–80000
5		Income	1	10000		0.9091	9091
6			2	25000		0.8264	20661
7			3	35000		0.7513	26296
8			4	30000		0.6830	20490
9			5	20000		0.6209	12418
10			NET PRESENT VALUE =				8957
11							

Note: The above assumes that the cells E3 is in a percentage format (see spreadsheet user 1); if not the values in the above formula require division by 100.

Your Net Present Value calculator spreadsheet should now look as shown on p 60 (over).
    The present value can be calculated directly from the cash flow and target rate using the NPV function.

The NPV function in Excel :   NPV(rate,range)   the net present value of a series of cashflows (range), at the discount rate

The NPV function in Lotus: @NPV(rate,range)   the net present value of a series of cashflows (range), at the discount rate

Note: in both NPV functions it is assumed that the first cash-flow occurs at the end of the first period, and subsequent cash-flows at the end of each subsequent period. In many property and other examples in reality the outlay will occur at the beginning of the first period as in this project example.

	A	B	C	D	E	F	G	H
1								
2			Year		P.V.£1@	P.V.£1@		
3					Discount Rate of	10.00%		
4		Expenditure		-80000			-80000	
5		Income	1	10000		0.9091	9091	
6			2	25000		0.8264	20661	
7			3	35000		0.7513	26296	
8			4	30000		0.6830	20490	
9			5	20000		0.6209	12418	
10					NET PRESENT VALUE =		8957	
11								
12								

Try calculating the NPV in Project 1 using the NPV function in a spare cell expressed as:

In Excel: =NPV (0.1,D4:D9)
In Lotus 1-2-3:   @NPV(0.1,D4:D9)

the figure produced is £8,143 : which differs from our calculation using the present value formula.

This undervaluation is caused because the NPV function is treating the outlay of £80,000 as occurring at the end of Period 0 not at the beginning.

To use the NPV function in Lotus or Excel it needs to be modified if, as in most cases, the initial outlay occurs at the beginning of period 0.

The initial cashflow must be isolated so the NPV function in Project 1 must be modified to:

In Excel: =-80000+NPV (0.1,D5:D9)
In Lotus 1-2-3:   -80000+@NPV(0.1,D5:D9)

and the same result as the original calculation using the present value formula should result.

Project 2: Calculating the IRR

The IRR function in both Lotus and Excel uses the iteration technique to calculate the discount rate at which the NPV for a given cash flow is equal to 0:

The IRR function in Excel :   =IRR(values, guess) the internal rate of return of a series of cash-flows (range), the guess rate being used only to start the iteration process.

The IRR function in Lotus :   @IRR(guess, range) the internal rate of return of a series ofcash-flows (range), the guess rate being used only to start the iteration process.

To calculate the IRR in Project 1 enter the following in a spare cell:

In Excel 4 :   = IRR (D4:D9,0.1)
In Lotus 1-2-3 :   @ IRR (0.1,D4:D9)

Note: for the IRR function to work there must usually be at least one positive and one negative value in the cash-flow. Problems may occur where the cash-flow is non-standard and it changes sign several times.

Add the Internal Rate of Return to your spreadsheet in Project 1:

	A	B	C	D	E	E	F
1							
2			YEAR			PV £1@	Discounted
3					Discount	10%	Sum
					Rate		
4		Expenditure		−80000			−80000
5		Income	1	10000		0.9091	9091
6			2	25000		0.8264	20661
7			3	35000		0.7513	26296
8			4	30000		0.6830	20490
9			5	20000		0.6209	12418
10			NET PRESENT VALUE =				8957
11							
12			INTERNAL RATE OF RETURN =				13.90%

To do this add the following formula to CELL G12:

= IRR(D4:D9,F3)

Your Net Present Value and Internal Rate of Return calculator spreadsheet should now look like this:

	A	B	C	D	E	F	G	H
1								
2			Year		P.V.£1@	P.V.£1@		
3					Discount Rate of	10.00%		
4		Expenditure		-80000			-80000	
5		Income	1	10000		0.9091	9091	
6			2	25000		0.8264	20661	
7			3	35000		0.7513	26296	
8			4	30000		0.6830	20490	
9			5	20000		0.6209	12418	
10					NET PRESENT VALUE =		8957	
11								
12					INTERNAL RATE OF RETURN =		13.90%	
13								
14								

Part One

Questions

1. A sum of £250 is invested at 9.5 % for 15 years.
   What sum will there be at the end of 15 years?

2. A capital sum of £1,500 is needed in 10 years. How much must be invested today assuming a compound rate of 8.5%?

3. A mortgage of £40,000 is arranged at 8%. What is the annual repayment if the term is for 30 years?

4. A sum of £1,000 is invested at the end of each year and earns interest at 16%. How much will this accumulate to in 10 years' time?

5. How much must be invested each year at 6% compound interest to produce £2,500 in 16 years' time?

6. You have expectations of becoming a partner in private practice in 10 years' time. This will cost you £25,000.

   (a) How could you provide for this?
   (b) How much would it cost at 12%?
   (c) How much would it cost today saving at 12%, if inflation is increasing the cost at 12%?

7. (a) If you save £200 a year for 10 years in a building society at 7%, how much will you have at the end of 10 years?
   (b) If you added a further sum of £2,000 at the beginning of year 5, how much would you have at the end of 10 years?
   (c) If you do (a) and (b) and leave it to accumulate for a further 5 years, how much will you have?

8. A shop is let at £10,000 a year for 5 years after which the rent will rise to £50,000 in perpetuity. What is its value on an 8% basis?

9. 8% Treasury Stock are selling at £80 per £100 face value certificate. What is the true rate of return if the stock has exactly 5 years to run to redemption?

Part 2

The Income Approach

Basic Concepts

Introduction

Valuation by direct capital comparison with sales in the market is the preferred method of valuation for most saleable goods and services. Valuation by this method is reliable provided that the sample of comparable sales is of sufficient size to draw realistic conclusions as to market conditions. This requires full knowledge of each transaction. Such a situation rarely exists in the market for investment property, and in the absence of directly comparable sales figures the valuer turns to the investment method. The investment method is used for valuing income-producing property whether freehold or leasehold, because as a method it most closely reflects the behaviour of the various parties operating in the property market.

Initially, the valuer considers the level of income or net benefits to be derived from the ownership of an interest in property, because investors are primarily concerned with the income and the risks attaching to that income when making investment decisions.

Valuation was earlier summarised as the estimation of the future benefits to be enjoyed from the ownership of a freehold or a leasehold interest in land or property, expressing those future benefits in terms of present worth. The valuer must therefore be able to assess these future net benefits (income) and be able to select the appropriate rate of interest in order to discount these benefits to derive their present value. The income approach to property valuation requires the valuer to concentrate on the assessment of the income pattern and the rate(s) of interest to be used to discount that income-flow.

Definitions

In a discipline that is derived from urban economics and investment analysis one would expect to find some common agreement as to the meaning of terms used by practitioners. This does not exist, so that additional problems may arise when advice is given to investors who are more acquainted with terms used by other financial advisers. The advice of the most expert valuer is of minimal value if it is misinterpreted, so a definition of terms used from hereon may be useful.

The explanations and definitions that follow may not achieve universal acceptance but are adhered to within this text and where indicated are mandatory on members of the Royal Institution of Chartered Surveyors (RICS) and Institute of Rating and Revenue Valuers (IRRV).

The question of value definition has bemused valuers and philosophers for centuries. It is recorded that Plato described 'the notion of value as the most difficult question of all sciences'. (Cited in Real Estate Appraising in Canada, The Appraisal Institute of Canada).

An early definition is set out in the 1938 edition of Mustoe, Eve and Ansteys *Complete Valuation Practice*:

> In ordinary speech the 'value' of a thing means the amount of money which that thing is worth in the open market. The valuer, whether he is valuing real property or chattels or livestock, endeavours to assess each article in terms of pounds, shillings and pence; to him 'value' means the amount of money for which property will exchange.

Considerable difficulties emerged in the 1970s and again in the 1990s, over the meaning and definition of value. The RICS addressed the issues through the Asset Valuation Standards Committee in the 1970s and reviewed the issue through a working party set up in 1993 under Michael Mallinson. Many of the recommendations of the Mallinson Report have been accepted by the RICS who now require valuers to use the following definition of Market Value (MV).

Market value

This definition is the one adopted by the RICS, it is an international definition and was developed by the International Valuation Standards Committee (IVSC).

> The estimated amount for which a property should exchange on the date of valuation between a willing buyer and a willing seller in an arm's-length transaction after proper marketing wherein the parties had each acted knowledgeably, prudently and without compulsion'(RICS Appraisal and Valuation Standards — the Red Book — Practice Statement (PS 3.2).

The definition is detailed in PS 3.2 and valuers must report to clients the fact that 'each element of the definition has its own conceptual framework' PS 3.2.1.

Value in exchange

This is a term used by economists and others and is commonly what the RICS defines as market value. If a good or service is incapable of being exchanged for other goods or services or money equivalent then it has no market value.

Worth

This is defined in the Glossary of Terms in the Red Book as

> The value of property to a particular investor, or class of investors, for identified investment objective, In this context an investor could include an owner-occupier.

The extent to which a calculation of worth, at the stated date, differs from MV should be set out clearly and under no circumstances should a calculation of worth be described as a valuation.

Price

Price is a historic fact except when qualified in such a phrase as 'offered at an asking price of ...'. Under perfect market conditions value in use would equate with value in exchange and price would be synonymous with value. The property market is not perfect and price and market value cannot always be said to be equal.

Valuation

This is the art or science of estimating the value of interests in property. According to the dictionary, the word 'valuation' is interchangeable with the word 'appraisal'. The latter word is used by Baum and Crosby (1995) as a source term covering both the assessment of market value and for the estimation of worth. The latter is better considered as analysis. In the market the term 'appraisal' is sometimes used in reference to the 'Appraisal of a Development Scheme' or as implying a process that is more comprehensive than a mere opinion of value. For clarity this text uses the terms valuation and analysis.

Valuation is defined by the RICS in the Appraisal and Valuation Standards as

A member's opinion of the value of a specified interest or interests, at the date of valuation, given in writing. Unless limitations are agreed in the Terms and Conditions this will be provided after an inspection, and any further investigations and enquiries that are appropriate, having regard to the nature of the property and the purpose of the valuation.

Valuation report

This is the formal presentation of the valuer's opinion in written form. As a minimum it must contain a sufficient description to identify the property without doubt; a value definition; a statement as to the interest being valued and any legal encumbrances; the effective date of the valuation; any special feature of the property or the market that the valuer has taken special note of; and the value estimate itself.

These definitions may be confusing to the reader and not obviously that important, but in practice they may be very important. Even more important is the valuer's instruction and the importance of communication between valuer and client. 'How much is it worth?', 'What price should I offer?', 'Is it worth £x million?', 'What figure should we include in our accounts for the value of our property assets?' and 'Is that the same as their market worth?', are questions which may give rise to different responses from the valuer and might result in the valuer expressing different opinions for the same property interest for different purposes. The base or definition must be clarified at the time of agreeing instructions and must be confirmed in the valuation report.

In terms of this book and the methods set out herein the terms 'value' or 'valuation' refer to the assessment of *Market Value* unless otherwise stated.

The investment method

The investment method is a method of estimating the present worth of the rights to future benefits to be derived from the ownership of a specific interest in a specific property under given market conditions. In property valuation these future rights can usually be expressed as future income (rent) and/or future reversionary capital value. The latter is in itself an expression of resale rights to future benefits. The process of converting future income flows to present value capital sums is known as capitalisation, which in essence is the summation of the future benefits each discounted to the present at an appropriate market-derived discount rate of interest. The terms *discount rate* and *capitalisation rate* are increasingly preferred to 'rate of interest', 'interest rate' or 'rate of return'. The use of the term 'rate of interest' should be restricted to borrowing, being the rate of interest charged on borrowed funds or the rate of interest to be earned in a savings account.

In Chapter 1 a distinction was made between return on capital and return of capital. The interest rate or rate of return refers to return on capital only and is sometimes referred to as the remunerative rate to distinguish it from the return of capital or sinking fund rate. Technically a discount rate ignores capital recovery but in general valuation usage discount rate and capitalisation rate are synonymous and are assumed to mean any annual percentage rate used to convert a future benefit (income flow or lump sum) into a present worth estimate.

It is necessary to discount future sums at a rate to overcome:

- liquidity preference
- time preference
- the risks associated with uncertainty about the future.

Some of these risks have been identified by Baum and Crosby in their authoritative text on Investment Valuation and include:

- Tenant risk   voids
                       non payment of rent
                       breach of covenants
- Sector risk   type specific eg retail, offices
                       location specific
- Structural risk   building failure
                       accelerated depreciation or obsolescence
- Legislation risk   new laws specific to property that might affect usability, rents, etc.
- Taxation risk   possibility of future taxes on property
- Planning risk   planning policy may impinge upon performance eg ... new roads,
                       pedestrianisation
- Legal risk   undiscovered issues affecting legal title eg rights of way, unregistered land

To which must be added the uncertainty associated with the macro economy and inflation. A capitalisation rate reflects all these factors and more and is referred to as an all risks yield (ARY or Cap rate).

In its simplest form income capitalisation is merely the division of income (*I*) by the annual rate of capitalisation (*R*).

A rate of capitalisation is expressed as a percentage — say 10% — but as was stated in Chapter 1 all financial analyses and calculations require this ratio to be expressed as a decimal.

Therefore:

10% becomes 0.10,
and 8% = 0.08, and thus
Value $(V) = I / i$

Example 5.1

Calculate the present worth of a freehold interest in a property expected to produce a rental of £10,000 pa in perpetuity at an annual rate of capitalisation of 10%

$$V = I / i = \frac{10{,}000}{0.10} = £100{,}000$$

In preference to dividing by i UK valuers have always used the reciprocal of i, and multiplied the Income I by a capitalisation factor which is called the Years' Purchase.

Thus:

$$V = I \times 1 / i = £10{,}000 \times \frac{1}{0.10} = £10{,}000 \times 10 = £100{,}000$$

The capitalisation of an infinite income, dividing by a market rate of capitalisation, is a simple exercise known as capitalisation in perpetuity. For many valuations this is all that is necessary and some valuers customarily create very simple tables to undertake investment valuations as closely as possible to a form of direct capital comparison. This is possible in defined markets where the yield variation will be in a narrow range and the market rental for that type of property in that locality is also known to be in a narrow range. This process is useful to establish an approximate level of value before undertaking the more detailed analysis and valuations described in the rest of this chapter.

For example office space in the central business area of a town may typically let at net rents of between £150 and £175 per m² and freehold investment sales may consistently occur at yields of between 8% and 9%. The combinations of value per m² can conveniently be presented using excel or set out to the nearest £ in a table.

Rent Yield	150	155	160	165	170	175
8%	1875	1937	2000	2062	2125	2187
8.25%	1818	1879	1939	2000	2060	2121
8.5%	1765	1823	1882	1941	2000	2059
8.75%	1714	1771	1828	1886	1943	2000
9%	1667	1722	1778	1833	1889	1944

In a relatively stable market such a table may be useable for a year or more. It also provides a useful comparative check against other properties and sales that have occurred. The valuation of a 1,500 m² office building now requires the valuer to assess, in their opinion, based on comparison with other known lettings and sales, the rent and correct yield — say £160 and 8.75% and the valuation becomes 1,500 × £1828 = £2,742,000, which the experienced valuer might round to £2,750,000. The trainee valuer may well be advised to become aware of ball park figures such as these as they provide a useful check when undertaking more complex investment valuations ie if the property being valued was currently let at £140 per m² with a rent review in two years to a potential market rent of £160 the value of the encumbered freehold (that is the freehold subject to the current lease arrangements) must be less than £2,750,000.

Where, however, the income is finite, then allowance must be made for recovery of capital. Two concepts exist in UK valuation practice: the internal rate of return and the sinking fund return. Each allows for the systematic return of capital over the life of the investment, but each approach is based on different assumptions.

The internal rate of return assumes that capital recovery is at the same rate as the return on capital *i*. It reflects the normal investment criterion of a return on capital outstanding from year to year and at risk, with the capital being returned from year to year out of income. Where capital recovery is at the same rate as the risk rate the table used is the present value of £1 pa (YP single rate) which in turn is the reciprocal of the annuity £1 will purchase. Thus any finite income stream can be treated as an annuity calculation. This concept is the more acceptable because the present worth of any future sum is that sum which if invested today would accumulate at compound interest to that future sum, and hence the present worth of a number of such sums is the sum of their present values.

If the sums are equal and receivable in arrears then

$$V \quad = \quad I \times \left(\frac{1 - PV}{i} \right)$$

If the sums vary from year to year then:

$$V \quad = \quad \sum \frac{I_1}{(1 + i)} \quad + \quad \frac{I_2}{(1 + i)^2} \quad + \quad \frac{I_3}{(1 + i)^3} \quad \dots \quad \frac{I_n}{(1 + i)^n}$$

Mathematically there need be no distinction between the two concepts as capital recovery can always be provided within a capitalisation factor by incorporating a sinking fund to recover capital:

$$\text{PV £1 pa} \quad = \quad \frac{1}{i + ASF}$$

Example 5.2

Calculate the value today of the right to receive an income of £1,000 at the end of each year for the next 5 years if a return of 10 per cent is to be received from the purchase.

Income	£1,000
PV of £1 pa for 5 years at 10% (Years' Purchase)	3.7908
	£3,790.80

Proof IRR basis:

Year	Capital	Income	Interest at 10%	Capital recovered
1	3,790.80	1,000	379.80	620.92
2	3,169.88	1,000	316.99	683.01
3	2,486.87	1,000	248.68	751.32
4	1,735.55	1,000	173.56	826.44
5	909.11	1,000	90.91	909.09

Proof SF basis:

Capital	£3,790.80
Annual Sinking Fund to replace £1 at 10% over 5 years =	0.16380
	£620.9333

Therefore:

Income net of capital recovery	=	£1,000 – £ 620.933
	=	£379.067 and
	=	(£379.067/£3790.8) × 100 = 10%

But if it can be shown that investors insist on a return on initial outlay throughout the life of the investment, then the sinking fund concept is the more acceptable. Further, if it can be shown that investors expect a return of capital at a different rate to the return on capital then this can be allowed for in the formula (see p27).

The arithmetic manipulation of figures in capitalisation exercises is not, of course, valuation. Valuation is a process which requires careful consideration of a number of variables before figures can be substituted in mathematically proven formulae. Any assessment of present worth or market value can only be as good as the data input allows and that factor is dependent upon the education, skill and market experience of the valuer. Ability to analyse and understand the market is of paramount importance. On this point and in the context of the discussion on internal rate of return and sinking fund rate of return it must again be noted that the only true analysis that can be made of an investment sale price is the establishment of the internal rate of return. The relationship between a price and a known finite income can only be analysed to assess the sinking fund return if a prior decision is made as to the rate of interest at which the sinking fund is to accumulate.

Valuation has been likened to a science, not because of any precision that may or may not exist, or because in part it involves certain basic mathematics, but because the question 'How much?' poses a problem that requires a solution.

The scientific approach to problem solving is to follow a systematic process. There may well be short cuts within the process. Indeed the discounting exercise itself, because it is repetitive, can frequently be carried out by pre-programmed calculators and computers. Short cuts exist if data are already available, but adoption of a systematic approach provides the confidence that full account has been taken of all the factors likely to affect the value of a property. This systematic approach is outlined in the Valuation Process.

Valuation process

First the valuer should define the problem.

- 'What are the client's real requirements?'
- 'Why does he/she want a valuation?'
- 'What is the purpose of the valuation?'

These questions should establish whether the client requires a market valuation, or a valuation for company asset purposes, insurance, or rating.

The date of the valuation must be ascertained. If a value is required for book purposes at a certain date, an 'in advance rental' may require an 'in arrears' valuation and vice versa.

Following the Mallinson Report the RICS has reconfirmed the need for the valuer to clarify and agree conditions of engagement; that is, to confirm the client's instructions and to do so in writing. This stage in the process is a mandatory requirement for RICS and IRRV members, the details are set out in the RICS Appraisal and Valuation Standards. Most students can readily access this through their libraries or learning centres on line. An additional resource for most surveying students is the on line RICS Books isurv valuation service (*isurv.co.uk*).

Having established the fact that the instructions are within the valuer's competence and having confirmed the instructions, the valuer needs to arrange to inspect the property and to collect all the data needed for the valuation.

Within specific economic and market conditions the valuer will be considering five principal qualities:

- the quality of the legal title
- the quality of the location
- the quality of the building
- the quality of the lease(s)
- the quality of the tenant(s) in occupation or those that can be assumed in the case of a new or owner-occupied property.

The following schedule and Figure 5.1 provide some idea of the mass of data that needs to be collected and assessed in order to arrive at an objective opinion of the property's marketable qualities' which in turn colour the valuer's judgement as to the income generating capabilities of the property and its suitability as an investment.

The property: site measurements
building measurements:
    external, internal, number of floors (In accordance with RICS code of Measurement Practice)
elevation, orientation
services:
    heating, lighting,
    air conditioning, lifts, etc.
age and design
energy rating
suitability of premises for present use
adaptability
accessibility to markets, amenities, labour

Legal: interest to be valued:
    freehold or leasehold
    details of title restrictions such as: restrictive covenants
    details of any leases or sub-leases, tenants, rent levels, lease terms

Planning: permitted uses

Economic: general
state of economy
regional and local
    population structure
average wages
    principal employment
    state of local industry
    economic base of area
    level of unemployment
    town and regional growth prospects
    transportation, existing and planned
    current planning proposals
    building societies, savings banks and general level of investment in the town and region
    position of town in regional hierarchy

Market: total stock of similar property
comparison of subject property to the stock
new stock in course of construction and planned
vacancy rates
general level of rents and rates
tenant demand for similar property

Alternative
investments: other properties
stocks and shares

Legislation: planning control
landlord and tenant control over rents and security of tenure
privity of contract
safety, health and working conditions and controls and compliance with fire and disability regulations
European Union Directives

This data is collected for analysis if considered significant in terms of value. The valuer needs to know what other properties are in the market, whether they are better in terms of location, etc, who are in the market as potential purchasers and as potential tenants, and whether the market for the subject property is active.

To collect the level of information required entails either very sound local knowledge, or the need to make enquiries of the local planning authorities, rate collecting authorities, highway authorities, transport companies and local census statistics.

Thus one finds that the major firms of surveyors and valuers will now hold on file considerable detailed information on the City and West End of London, including in some instances very detailed street-by-street information, and similar information on the main provincial cities. This information source is kept up to date by research members extracting relevant information from national and local papers, local council minutes, etc. There are now many online sources of economic, market and general data relevant to real estate including access to data on property transactions held by the Land

---

Factors determining investors' yield requirements and market capitalisation rates

**General level
of interest rates**
*Economy / Government
Policy / Alternative
Investment Opportunities*

Legislation

*Planning / Landlord & Tenant / Building
 Health / Fire / Disability
Working Conditions
Environmental / EU*

Age and condition of building
*Repairs / Fashion /
Adaptability*

Inflation
*Rental growth
Capital growth*

Location
*Accessibility
Markets Labour*

Security
*Tenant / Rent
Lease Terms*

*Taxation
Rates / Income / Corporation / CGT / CTT*

Liquidity
Ease / Cost of transfer

Volatility
*Sensitivity to market
Change*

Management
*Tenant Mix / Service Charges
Lease Responsibilities*

Figure 5.1

Registry. Issues of due diligence means that for many valuers ignorance of such data and failure to make appropriate use of any such relevant data, if it lead to a negligent valuation, could result in action by clients for negligence.

Having collected the information shown above, it is then necessary to consider the market.

> A major danger in assessing the direct property market lies in failing to identify the main segments into which the market is divided according to the value, reversionary terms, etc ... of a property

This comment was made in the Greenwell & Co Property Report in October 1976 but is as true today as it was 30 years ago.

The valuer must be able to identify:

* the most probable type of purchaser
* the alternative comparable properties on the market for that purchaser market

---

The Valuation Process
Assess gross income (rent) from lease or estimate gross income (rent) potential from the market

Deduct
Allowances for voids

equals
Effective gross income

Deduct
Allowances for outgoings on repairs, insurance, management and any other operating expenses or non-recoverable service costs including rates

equals

Implied net income   *Contracted net income*

Capitalised (with reversion if appropriate) to produce estimate of total present value

---

Figure 5.2

- the current level of demand within that sub market and
- any new construction work in hand which will offer better/newer and, today, energy efficient space, for tenants which will impact on investor demand for existing properties.

These and other questions will indicate what market data is required as preparatory material for the valuation. A particular concern to valuers is the need to identify the possibility of the property containing deleterious materials and/or of the site or buildings being contaminated. The valuer needs to alert the client to the possibility that the valuation report will contain a number of exclusions or caveats, the recommended caveats are contained in the RICS Practice Statements.

The most difficult aspect in an income capitalisation exercise is the determination of the correct capitalisation rate. Every property investment is different, and if the available data on sales are insufficiently comparable then it may become difficult to justify the use of a selected rate. Thus if a property with a five year rent review pattern sells on a 7% basis then this can be assumed to be the market capitalisation rate for that type of property let on five year review pattern with the first review at a comparable date in the future and with a comparable level of rental increase.

If the subject property differs in any respect then the valuer will need to seek better comparables or will have to adjust that rate to reflect the differences between the comparable property and the subject property.

The valuation process can be redefined as shown Figure 5.2.

Within this process the three main variables likely to have the greatest effect on the final estimate of value can be identified: income (rent), operating expenses and capitalisation rate. It is our presumption that a valuer practicing in a well-run valuation department will be able to provide reliable figures for these items readily, and should be able to substantiate these figures from available analysed data.

Income or rent

Capitalisation is the expression of future benefits in terms of their present value. Valuation therefore requires the valuer to consider the future: but current UK valuation practice reflects a distrust of making predictions. The convention has developed of using initial yields on rack rented property as the capitalisation rates to be applied to current estimates of future rental income, thereby building into the capitalisation rate the market's forecast of future expectations. The forecast is still made but the valuer has avoided any explicit statement. American appraisers consider this approach to be outmoded and argue that it should only be used if a level constant income flow is the most probable income pattern. In other circumstances they recommend the use of a Discounted Cash Flow Approach.

The RICS research report (Trott, 1980) on valuation methods recommended the use of growth explicit models for investment analysis and suggested that a greater use should be made of growth explicit DCF models in the market valuation of property investments.

This view is supported by Baum and Crosby(1995), and the RICS Books publication *Commercial Investment Property: Valuation Methods,* An Information Paper.

Whenever the investment method is to be used, whether in relation to tenant-occupied property or owner-occupied property, an initial essential step is the estimation of the current rental value of that property. If the property is already let this allows the valuer to consider objectively the nature of the present rent roll. Initially the most important task must be the estimation of rental income.

If the property is owner-occupied then it is necessary to assess the imputed rental income. The assessment requires analysis of current rents being paid, rents being quoted and the vacancy rate of comparable properties.

This assessment should be carried out in accordance with the RICS definition of Market Rent (MR) set out in the Red Book as follows:

> The estimated amount for which a property, or space within a property, should lease(let) on the date of valuation between a willing lessor and a willing lessee on appropriate terms, in an arm's-length transaction, after proper marketing wherein the parties had each acted knowledgeably, prudently and without compulsion.

Again 'this definition must be applied in accordance with the conceptual framework of MV at PS3.2 together with the ... supplementary commentary' (PS3.4).

It will be seen later that this differs to the statutory definition in the Landlord and Tenant Act 1954 and will differ to many sets of assumptions set out in existing leases. In assessing the rental value of a property the valuer needs to check the basis for such calculations before turning to the market for supporting evidence see Figure 5.3.

Detailed analysis of the letting market must precede analysis of a specific letting. Thus if it can be demonstrated within the market for office space that such space users will pay a higher rent for ground floor space than for space on a higher floor then this may be reflected in the way in which a letting of a whole office building is analysed in terms of rent per m². If the rent of luxury flats and apartments can be shown to have a closer correlation with the number of bedrooms than with floor area then analysis might be possible on a per bedroom basis.

If floor area is to be used then continuing analysis will indicate whether or not bids are influenced by, for example:

- the size, quality and location of:
  — entrances
  — common areas, (providing toilet, vending and catering facilities)

Situation:	Basis:
Lease renewals of business premises under Landlord and Tenant Act 1954	section 34 Landlord and Tenant Act 1954
Rent reviews	Rent review clause within current lease
Vacant property, owner occupied property or where a reversion to a new letting can be assumed	RICS definition of MR unless otherwise agreed
Estimates of MR	RICS definition of MR

Figure 5.3 Rental basis

- the provision of:
  - air conditioning
  - car parking
  - security
  - accessibility (in some cases to facilitate 24/7 working practices)
  - lifts

- communal facilities:
    recreation, leisure, food
  - banks, cash points
  - first aid

- design and layout:
  - floor plate and structural grid layout
  - flexibility

- other matters:
  - energy efficiency
  - increasingly the provision of health and well being facilities including gyms, and relaxation areas.

Reliable estimates of current rents can only flow from analysis of rents actually being paid, and under no circumstances should a valuation be based on an estimate of rent derived from analysis of an investment sale where there is no current rack rent passing. This is because rent analysed in this manner depends upon the assumption of a capitalisation rate.

For example, if the only information available on a transaction is that a building of 500 m² has just been sold for £50,000 with vacant possession, the only possible analysis is of sale price per m².

In most cases analyses of rent being paid should be on the basis of net lettable space, but agricultural land is often quoted on a total area basis inclusive of farm dwellings and buildings.

Net lettable space becomes more meaningful if it is taken to mean the net area of the building suitable for use for the purpose let as defined in the RICS Code of Measurement. This definition excludes circulation space within the building such as stairwells, landings, lifts and ancillary facilities such as wash rooms. The rent analysed on this basis reflects the quantity and quality of the facilities provided, which should of course be noted on any data record sheet for future use. The precise set of rules for measuring buildings of different use types will vary from valuer to valuer and from area to area, but the old adage 'as you analyse so should you value' should be adhered to in terms of building measurement. Not only is it critical to know what should be included as part of the lettable area but the measurements themselves must be taken accurately.

There is a growing tendency for shop premises to be let at a fixed rental plus a percentage of profit or turnover. Where a valuation is required of property let on such terms full details of total rents actually collected, checked against audited accounts for a minimum of three years, should be used as the basis for determining income cash flows for valuation purposes.

Due to the heterogeneous nature of property it is customary to express rent in terms of a suitable unit of comparison thus:

*   agricultural land   rent per hectare
*   office and factory premises   rent per m^2
*   shops   rent per m^2 overall or per m^2 front zone

The measurement of retail premises: zoning

Three alternative approaches have been developed for analysing shop rent:

*   overall analysis
*   arithmetic zoning and
*   natural zoning.

Overall analysis

The rent for the retail space is divided by the lettable rental space to obtain an overall rent per m^2. This is a simple approach but is complicated by the practice of letting retail space together with space on upper floors used for storage, sales, rest rooms, offices or residential accommodation at a single rental figure. In these circumstances it is desirable to isolate the rent for the retail space.

Some valuers suggest there is a relationship between ground floor space and space on the upper floors — this will only be the case where the user is the same or ancillary (eg storage). Here custom or thorough analysis will indicate the relationship, if any, between ground floor and upper floor rental values. Where the use is different the rent of the upper floor should be assessed by comparison with similar space elsewhere and deducted from the total rent before analysis. Thus if the upper floors comprise flats then the rent for these should be assessed by comparison within the residential market and then deducted from the total rent.

Overall analysis tends to be used for shops in small parades and for large space users. In the latter case it is reasonable to argue that tenants of such premises will pay a pro rata rent for every additional m^2 up to a given maximum. The problem here is that what one retailer might consider to be a desirable maximum could be excessive for all other retailers. Such a point should be reflected as a risk accounted for by the valuer in the consideration of a capitalisation rate.

Arithmetic zoning

This approach is preferred in many cases to an overall rent analysis because, in retailing, it is the space used to attract the customer into the premises that is the most valuable (namely the frontage to the street or mall). Again the rent for the retail space should be isolated before analysis.

Example 5.3

Analyse the rent of £50,000 being paid for shop premises with a frontage of 6m and a depth of 21m.

Overall $= £50,000 \div 126(6 \times 21) = £396.82$ per m^2

Zoning: Assume zones of 7m depth and £x per m^2 rent for Zone A

Then	Zone A	$= 6 \times 7 =$	42 m^2 at £x	$=$	42.0x
	Zone B	$= 6 \times 7 =$	42 m^2 at £0.5x	$=$	21.0x
	Zone C	$= 6 \times 7 =$	42 m^2 at £0.25x	$=$	10.5x
					73.5x

$£50,000 \div 73.5x = £680.27$

The rental value for Zone A is therefore £680 per m^2.

The observant reader will realise that the space could be divided into any number of zones. Different regions and different retail centres display different trading patterns, and while a common convention is to use two zones and a remainder with Zone A and B of 7m depth other practices will be found. Custom and practice do not necessarily reflect market behaviour and adapting different zones and zone depths for analysis can produce considerable variations in opinion as to rental value for a given location. A continuing concern is the switch from imperial measures where 20 ft zones or their equivalent metric figure may be replaced by depths expressed initially in metres eg 6m/7m Zone A depths. The arithmetical implications on rental analysis and hence opinions of rental value are very real; but rents do not vary because of variations in the zones used by valuers, they vary because of variations in trading potential.

Certainly in practice retailers do not see the premises divided into rigid zones. Every rental estimate must be looked at in the light of the current market and common sense — who would be the most probable tenant for premises 7m wide by 100m deep? Is there any user who operates in such space, or is the last 50m waste or valueless space for most retailers in that locality?

Natural zoning

This method can only effectively be used to analyse rents within a shopping street or centre where information is available on a number of units, as it requires comparison between units. As previously explained the rents for retail space must be isolated from the rent for the premises as a whole. It assumes that there is a base rent payable for a given area of space and that the extra rent paid for larger areas represents the additional rent for the extra floor area. It is difficult to use as an ongoing basis of analysis. This technique was originally detailed in R Emeny and HM Wilks,'Principles and Practice of Rating Valuation', Estates Gazette, 1982.

The rent of two adjoining premises, one of 6m × 21m, let at £50,000 pa, the other 8m × 25m, let at £65,000 pa can then be considered.

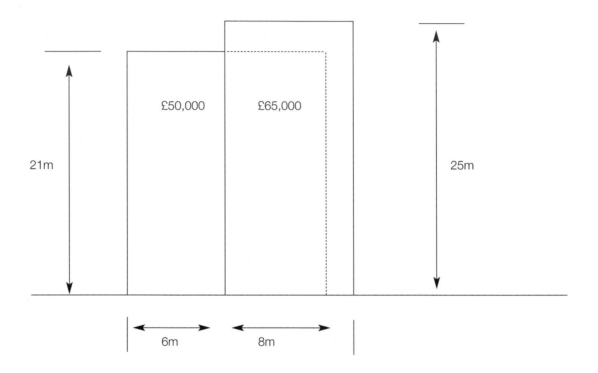

The method here argues that if retailers are paying £50,000 for 6m × 21m and £65,000 for 8m × 25m then £15,000 represents the rent for the additional space shown in the sketch. Even here the logic may collapse if somewhere within the parade there is a shop 6m × 20m let at £67,500 pa.

In terms of the rental analysis of retail space, warehouse accommodation and factory space, the valuer needs to have a sound knowledge of the specific requirements of different retailers and of different manufacturers.

Where one is dealing with standard shop units, for example within a shopping mall, the unit of comparison can often be left as the 'shop unit'. Rent will be a factor of location/position and not size.

The developing technique, although not yet found to be any more satisfactory than those listed, is multivariate analysis when the dependent variable rent is considered against a number of independent variables which could include size, location, distance from car parks, bus station, etc.

It would be ideal if all rental estimates could be based on true comparables, ie those of the same size, design, facilities, location, etc. This is rarely the case in practice. With experience, adjustments can be made for some of these variations, but wherever possible estimates of rent should be based on close comparables.

As far as office records are concerned, strict procedures should be adhered to so that the format of data is consistent. In this respect it is recommended that all rents are analysed net of landlord's outgoings (see below) and that apart from obvious factors such as the address of the property the record should contain details of facilities included (eg central heating, air conditioning) and the lease terms.

The valuer must be in possession of all the facts of a given letting before any rental figure can be analysed. In the case of residential and commercial property statute intervenes to protect the tenant in several ways (see Chapter 10). Thus a tenant of business premises who carries out improvements would not expect to pay an increase in rent for these improvements on the renewal of the lease. The valuer must therefore know the basis upon which a given rent was agreed before he can make proper use of the information. An earlier lease might have been surrendered or the tenant might have undertaken to modernise the premises. Either of these could have resulted in a rental lower than a market rental being agreed between the parties. And of increasing concern is the lease definition of rent where there is provision for a rent review — there is almost certainly a difference between 'full market rental' and a 'reasonable rent', the latter inferring something less than the maximum rent achievable if offered in the open market.

The RICS has advised that where rental evidence is obtained from other valuers it should be provided on a consistent basis.

It is arguable that the rent agreed for a 20-year lease without review will differ from that agreed for the same lease with 10-year, seven-year or five-year reviews. Whether these rents would be higher or lower will depend upon whether the rental market is rising or falling (see Chapter 9). For preference, comparability should entail estimating rents from analyses of lettings with similar review patterns. Where this is not possible the valuer will have to exercise professional judgment in making adjustments from the comparable evidence before using the evidence to estimate current market value of the subject property.

The need in a weak letting market for landlords to offer inducements such as fitting out costs and rent-free periods indicates the need for valuers to be in full possession of all the facts before reaching any conclusion on market rents being paid.

Crudely a rent of £50,000 pa agreed for a property for a period of five years with no rent payable for one year would be viewed by the market as a rent of £40,000 pa.

(£50,000 × 4 years) divided by 5 years = £40,000

The landlord's argument might be that the rental value is £50,000 a year but it has been waived for 12 months suggesting a philanthropic landlord. But such a genuine arrangement is possible and the valuer needs to know the full details of all rents being analysed or used as evidence. The reality is that the market will be aware of the general tone of rents at a given point in time and most valuers will know whether the £50,000 is a 'headline rent' or close to the market levels.

Other inducements can be considered in a similar way. Thus £100,000 given for fitting out the building might be treated as equivalent over five years to a reduction of £20,000 a year, ie, the rental value without the £100,000 would have had to fall to £30,000 a year to induce a letting.

The main issue with all such inducements is assessing the period of time to which the inducement applies. The authors' opinion is that it should apply only to the period up to the first full rent review and, strictly speaking a discounting approach should be adopted to reflect the time value of money. (See RICS Valuation Information Paper No 8 The analysis of commercial lease transactions, 2006.)

Rental value will also be affected by a number of other lease clauses. Particular attention should be paid to those dealing with alienation — assignment and subletting; user; repair and service charges.

Although the Landlord and Tenant Act 1988 now ensures that landlords do not unduly delay the granting of consents in conditional cases, the position in relation to an absolute prohibition on alienation remains unchanged. In the latter case there will be an adverse effect on rents. User restrictions in shopping centres and multiple tenanted commercial property can have a beneficial effect on rents, but where they are too restrictive they will have a detrimental effect on rents.

The Landlord and Tenant (Covenants) Act 1995 has amended the law on Privity of Contract. It allows landlords to define in a lease the qualifications of acceptable assignees and whilst 'reasonableness' remains a factor, some leases will be seen to be more onerous than others due to such legally permitted restrictions.

Responsibility for repairs and the nature of any service charge provisions must be considered twice: first to see if they are affecting market rents, and second to see if they leave the landlord with a liability which must be estimated and deducted as a landlord's expense before capitalisation. (For valuation purposes rent must be net of VAT.)

Landlords' expenses

Where investment property is already subject to a contracted lease rent, and/or where it is customary to quote rents for a specific type of property on gross terms, it is essential to deduct landlords' outgoings (operating expenses) before capitalising the income. Investment valuations must be based on net income. This means an income net of all the expenses that the owner of an interest in property is required to meet out of the rents received from ownership of that interest in that property other than tax.

Expenses may be imposed upon the landlord by legislation, or they may be contractual, as in an existing lease. But an inspection of the property may suggest that, though neither party is statutorily or contractually liable, there are other expenses that will have to be met by the owner of that interest in the property. The valuer must identify all such liabilities and make full allowance for them in the valuation. In order to do this, reference must be made to all existing leases in respect of the property.

The principal items of expenditure can be broadly classified under the headings of insurance, management, taxes, running expenses and repairs. Of these only the cost of complying with repairing obligations should cause any real difficulty in accurate assessment.

Insurance

The valuer armed with plans and his own detailed measurements of the building will be able to estimate or obtain an accurate quotation for all insurances, particularly fire insurance. Fire insurance is an extremely involved subject, complicated by the variation in insurance policies offered. A valuer should therefore be acquainted with the terms and conditions of policies offered by two or three leading insurance companies and should always assess the insurance premium in accordance with those policies. If reinstatement value is required this can be referred to a building surveyor or preferably a quantity surveyor, or may be based on adjusted average figures extracted from Spon's *Architects' and Builders' Price Book.* Due to the 'averaging provisions' of most policies it is better to be over-insured than under-insured, and hence to overestimate rather than underestimate this item.

Quotations for other insurances necessary on boilers, lifts, etc, can always be obtained from a broker or insurance company.

A deduction for insurance will rarely be necessary as the lease will usually contain provisions for the recovery of all insurance charges from the tenant in addition to the rent. In most cases the wording of a 'full repairing and insuring lease' leaves the responsibility for insurance with the owner, but the cost of the insurance with the tenant.

Management

This refers to the property owner's supervising costs equivalent to the fee that would be due to a management agent for rent collection; attendance to day-to-day matters such as the granting of licences to assign/sublet or to alter the premises; inspections of the premises, and instructions to builders to carry out repairs. It is frequently considered to be too small to allow for in the case of premises let to first class tenants on full repairing and insuring terms. The valuer must use discretion but it is suggested that if the valuer's firm would charge a fee for acting as managing agents, then a similar cost will be incurred by any owner of the property and a deduction for management should therefore be made.

The fee may be based on a percentage of rents and service charges collected or on a negotiated annual fee. The practising valuer will be aware of the appropriate market adjustments to make, based on current fees charged by managing agents. Where VAT is payable on rents it will impose an additional management cost which will be reflected in the fees negotiated by the managing agents. Where the management fee is recoverable as part of a full service charge no deduction need be made in a valuation of the landlord's interest.

Taxes

The Uniform Business Rate (UBR) is paid by the occupier of business premises. Inspection of the lease will indicate the party to that lease who has contracted to meet the 'rates' demand. If premises are let at an inclusive rental (ie inclusive of UBR) then rates should be deducted. If let exclusive of UBR then no deduction is necessary. If the letting is inclusive and there is no 'excess rates clause' then the deduction must represent the average annual figure expected up until the end of the lease or next review, not a figure representing current rates. If an excess rates clause is included then the sum to be deducted is the amount of rates due in the first year of the lease or the lowest sum demanded during the current lease. This is because such a clause allows for the recovery from the tenant of any increase in rates over and above the amount due in the first year of the lease.

Other 'rates' may include water rates, drainage rates, and rates for environmental and other purposes.

It has always been the custom in the UK to value before deduction of tax on the grounds that income and corporation tax are related to the individual or company, and not to the property. However, if market value implies the most probable selling price, there is automatically an implication of the most probable purchaser. The valuer without this knowledge cannot be held to be assessing market value; thus it is held by some that the tax liability of the most probable purchaser should be reflected in the valuation (see Chapter 8).

For simplicity, no deduction will be made for tax in most of this text.

Running expenses

Where the owner of an interest is responsible for the day to day running of a property such as a block of flats, and office building let in suites, or a modern shopping centre, a deduction from rent may be necessary to cover the cost of items such as heating, lighting, cleaning and porterage. Current practice is to include provision for a separate service charge to be levied to cover the full cost of most running expenses.

The valuer must therefore inspect the leases to check the extent to which such expenses are recoverable. Older leases tend to include partial service charges, in which case the total income from the property should be assessed and the total cost of services deducted.

All the items falling under the general head of 'running expenses' are capable of accurate assessment by reference to current accounts for the subject property; by comparison with other properties; by enquiry of electricity, gas and oil suppliers; by enquiry of staff agencies; and so on. The valuer operating within a firm with a large management department is at a distinct advantage, as that department should be able to provide fairly accurate estimates based on detailed analyses of comparable managed property.

Where a full service charge is payable by the tenant this is usually adjustable in arrears. In other words, the annual service charge is based on last year's expenses. During periods of rapidly rising costs the owner will have to meet the difference between service cost and service charge, and reduction in the owner's cash flow and should be taken into account in a valuation.

Increasingly these points are being met by more complex service charge clauses and schedules which provide for interim increases; for example to cover the uncertain energy element in running expenses. (The statutory requirements relating to service charges and management charges in residential property are complex. Readers involved or interested in the residential sector must refer to all current Housing and Landlord and Tenant legislation.)

Repairs

A detailed consideration of existing leases will indicate those items of repair that have to be met by the landlord out of rental income. The sum to be deducted from an annual income before capitalisation must be an annual averaged figure. Thus liability to redecorate a building every five years should be estimated and averaged over the five years. A check should be made against double accounting — if the cost of repairs to boilers, lifts, etc, is covered by a service charge then no allowance should appear under repairs for such items. If indeed the cost of redecoration is recoverable by a direct proportionate charge to tenants then no deduction need be made.

Where an allowance has to be made every effort is required to estimate the amount as accurately as possible. An excess allowance will lead to an undervaluation, inadequate allowance to an overvaluation.

Advice, if needed, should be obtained from builders and building surveyors as well as by comparison with other known repair costs for comparable managed property.

Essential works

Any obvious immediate renewals or repairs as at the date of the valuation should be allowed for by a deduction from the estimate of capital value to reflect the cost (see Chapter 9).

Averages and percentages

Valuers should use averages with care. An example of this is the use of average heating costs per m². The valuer is required to value a specific property, not an average property. The requirement is to estimate the average annual cost of heating that specific property. Some valuers, texts, and correspondence courses suggest that it is a reasonable approach to base insurance premiums and repair costs on a percentage of market rental value. This approach is not recommended unless the valuer can prove that the figure adopted is correct. A percentage allowance may be widely inaccurate; a High Street shop could be let at £100,000 a year and a factory could be let at £100,000 a year, but the latter will almost certainly cost more to keep in repair. Two properties may let at identical rents, but one may be constructed of maintenance-free materials, the other more cheaply constructed with short-life material. A thousand m^2 of space in Oxford Street, London, and in Exeter, Devon, may cost the same to maintain but could have very different market rental values. A large old building may let at the same rent as a small modern building, and clearly the repairs will differ.

Similar points may be made in respect of the use of percentages to estimate insurance premiums.

Voids

Where there is an over-supply of space within a given area the probability of voids occurring when leases terminate is increased. If a single investment building is let in suites to n tenants, voids may occur sufficiently regularly for a valuer to conclude that the average occupancy is only, say, 90%. In such cases, having estimated the market rental value, the figure should be reduced to the level of the most probable annual amount. Voids are less likely to affect operating expenses so a pro rata allowance on operating costs should not be made. Indeed in some exceptional cases it may have to be increased to allow for empty rates, non-recoverable service costs and additional security.

In the case of a building let to a single tenant, a void is only likely to occur at the end of the current lease. If this is a reasonable expectation then the income pattern must be assumed to be broken at the renewal date for the length of time considered necessary to allow for finding a new tenant. Additional allowances may have to be made for negative cash flows if empty rates and other running expenses have to be met by the owner during this period.

The time taken to renew leases to current tenants and to negotiate rent reviews can represent a significant loss to a landlord. New leases will contain 'interest' clauses to cover interest lost on rent arrears, and delayed settlement of reviews and renewals.

In other cases the valuer may wish to adjust rental income down to reflect the fact that agreement is unlikely to coincide with the rent review or lease renewal date.

Purchase expenses

Stamp Duty Land Tax, solicitors' fees, etc

It has long been valuation practice when giving investment advice to include in the total purchase price an allowance to cover stamp duty, solicitors' fees, and any other expenses to the purchaser occasioned by the transaction including VAT. This in total currently amounts for most property to 5.75% of the purchase price.

It should be remembered that analysis of a transaction might also take these expenses into account to reveal the purchaser's yield on total outlay rather than the yield realised on the purchase price alone.

It is current practice in investment valuation work to express value net of these sums. However, MV is normally expressed without deduction for selling expenses.

Income capitalisation or DCF

The first step in the income approach is to identify the actual net rent being paid (Net Operating Income: NOI in North American appraisal practice), and the current market rent in net terms.

The second step is to determine the most appropriate basis of discounting the income; that is, whether to use:

- non-growth Discounted Cash Flow
- growth explicit Discounted Cash Flow
- conventional income capitalisation
- rational or real value methods.

In an active property market the valuer will be able to determine an appropriate capitalisation rate (all risks yield) from the analysis of sales of comparable properties.

For a property to be comparable it will need to be:

- similar in use
- similar in location
- similar in age, condition, design, etc (ie to have similar utility value)
- similar in quality of tenant
- let on similar lease terms.

Where a property sale meets most but not all of these criteria the experienced valuer with good market knowledge will be able to adjust the capitalisation evidence obtained from the market for any variations.

Experience or knowledge here means knowing:

- the current players in the market
- who is selling which type of property and why
- who is buying which type of property and why
- the current state of the market and the underlying reasons for it rising or falling
- the national, regional and local economic situation.

Analysis of current sales prices must be undertaken on a consistent basis, the valuer needs full details of the transaction in order to assess the capitalisation rate net of purchase costs.

Example 5.4

An office building is currently let at its open market rent of £100,000 a year net of all outgoings with rent reviews every five years. It has just been sold for £1,250,000.

Assess the capitalisation rate.

$$\frac{100,000}{1,250,000} \times 100 = 8\%$$

In practice the transaction may have been completed at a price to produce a return of 8% on total acquisition cost to the purchaser and thus £1,250,000 would represent price plus fees hence:

let purchase price	=	£x
let purchase costs	=	5.75%
then 1.0575x	=	£1,250,000
x	=	£1,182,033
and capitalisation rate	=	8.46%

A variation in capitalisation rate of say 0.50% can be significant at the lower rates. Thus:

Income	£100,000
YP perp at 4.50%	22.22
	£2,352,941

Income	£100,000
YP perp at 4.00%	25.0
	£2,500,000

Here the 0.50% variation represents £278,000 or a variation in value of 11.12%.

The market evidence and market knowledge will allow the valuer to express an opinion on capitalisation rates with some confidence and to be able to make 'intuitive' or professional adjustments for variation between the subject property and the market evidence. If the market evidence is 8% and the valuer is to use that to value a comparable but marginally superior investment property the capitalisation rate will be adjusted down to less than 8%, if the property being valued is marginally inferior the rate will be adjusted to more than 8%. There is no rule but on a point of principle one might argue that any adjustment between comparable and subject property in excess of 0.25% might suggest the comparable is not comparable.

In some cases the property will display qualities which add to and detract from the quality of the investment. Thus a property may be in a marginally better location but may be marginally less efficient having only 75% net to gross floor area compared to the normal 80% for the market and so the adjustments cancel out.

The subtleties of market intuition cannot be taught, they have to be acquired from experience.

Where the property sold is producing a rent (income) below market rental value, and there is an expectation of an early reversion, the capitalisation rate can be found by trial and error or iteratively using a financial calculator.

This is an IRR calculation (see p52) and the rate percent found is called an equivalent yield (see p253).

Question 1

A freehold shop let at its market rental value of £75,000 a year with rent reviews every five years has just been sold for £1,250,000. Assess the market capitalisation rate.

Question 2

A freehold office let at £30,000 a year with 2 years to the next rent review has just been sold for £553,000. The current market rent is £40,000. Assess the equivalent yield by trial and error.

Discounted Cash Flow (DCF)

All income capitalisations are a simplified form of DCF. Thus the capitalisation of £100,000 a year in perpetuity at 8% can be shown to be the same as discounting a perpetual (for ever) cash flow of £100,000 receivable at the end of each year in perpetuity.

Nevertheless there are growing arguments for using a cash flow approach for all income valuations — in particular a DCF approach. This

- provides clearer meaning to investors, bankers and professional advisers
- can incorporate regular and irregular costs such as fees for rent reviews, lease renewals, voids, etc
- can allow for expected refurbishment costs
- can incorporate adjustments for obsolescence.

DCF is the normal basis in North America where property is frequently let on gross rents with provision for landlords to recoup increases in certain service costs but not all.

Additionally, DCF methods may be a preferred approach when dealing with new developments which are not expected to achieve full potential for several years following practical completion.

The current use of the term 'DCF' by UK valuers generally refers to the use of a modified DCF, (Rational Method) or Real Value approach to valuation. Some valuers believe these methods should be used for all valuations whilst others see their use being reserved for those properties displaying an income pattern which is out of line with the normal expectations in the marketplace.

The use of DCF and conventional capitalisation approaches are considered in Chapters 6 and 7 for the valuation of freehold and leasehold interests in property.

Summary

The key steps in the income approach to property valuation are:

- agree purpose of valuation, basis of value and other terms and conditions with the client
- assess current net operating income
- assess income potential based on comparable evidence of current lettings
- assess appropriate capitalisation rate
- consider use of cash flow layout in order to incorporate specific allowance for fees, depreciation, etc
- complete the valuation and prepare a full report
- in any valuation falling within the requirements of the RICS Appraisal and Valuation Standards the valuer, if a member of the RICS or IRRV, must comply with the mandatory standards.

Freeholds

The main legal interests bought and sold in property are freehold and leasehold interests. As investments the former have an assumed perpetual life because the freehold title (Fee simple absolute) is the largest legal estate or right of ownership of land in the UK, a title which enjoys rights in perpetuity; the latter have a limited life span fixed by the lease term. Having identified the interest to be valued the valuer must then consider what future benefits, by way of rent, ownership will produce for an investor and assess any liabilities that the owner may be, or may become, liable for in the future. This will usually involve assessing the market rent, determining what rent if any is currently being paid under any existing leases together with an estimate of any annual expenditures payable by the investor and any immediate or future capital liabilities. The timing of these potential receipts and payments must also be ascertained. The perpetual nature of freehold interests means that valuers have developed certain short cuts as clearly it is not sensible to try and look too far into the future.

Two distinct approaches have developed for assessing the market value of an income-producing property.

The first, as already explained in Chapter 4, is to assume a level continuous income flow and to use an overall or all-risks capitalisation rate derived from the analysis of sales of comparable properties let on similar terms and conditions (ie five year or seven year rent review patterns) to calculate present value, that is market value.

The second has been named the discounted cash flow (DCF) approach.

The first or normal approach is favoured by many on the grounds that it is more correctly a market valuation. As such it relies on an active property market and an ability to analyse and obtain details of comparable capitalisation rates. The definition of market value implies that there must be a market for the asset being valued. If not then the necessary ability to compare prices achieved to support an opinion of value is lost, there is no market so there can be no market value. However those in favour of the DCF approach argue that, when there is limited market activity, a valuation can still be undertaken using a DCF approach. Market activity in this respect refers not only to the total volume of sales at a point in time but a sufficiency of direct comparables. It will be shown later that at a time when there might not be any evidence by direct comparison or, in the case of the income approach, indirect comparison a value may be derived indirectly from other market activity and developed through the DCF approach.

For example, it is assumed that if initial yields on market rented properties are, say, 7% then the capitalisation rates to be used for valuing property with an early reversion to a market rent can be closely related to 7%. Further that if there is a long period of time to the reversion then the 7% can simply be adjusted upwards for the change in risk by the valuer. The argument against this is that if there is no evidence of capitalisation rates from sales of investments with early reversions or long reversions it may be because there is no market for such investments but in order to support an opinion of market value, required by a client, the DCF approach which is built up from the known market evidence is more supportable and likely to be more accurate than any intuitive adjustment made by a valuer.

The strongest criticisms of the normal approach are that it fails to specify explicitly the income flows and patterns assumed by the valuer and, that growth implicit all risk yields are used to capitalise fixed flows of income. The DCF approach requires the valuer to specify precisely what rental income and expenses are expected when, and for how long. The valuer is therefore forced to concentrate on the national and local economic issues likely to affect the value of the specific property as an investment. There may after all be properties in a depressed economy for which rental increases in the foreseeable future are very unlikely.

The DCF approach accepts the idea of the opportunity cost of investment funds. Opportunity cost implies that a rate of return must be paid to an investor sufficient to meet the competition of alternative investment outlets for the investor's funds. This is the basis of the risk free rate of discount, a risk free rate being assumed to compensate for time preference only. Any investment with a poorer liquidity factor or higher risk to income or capital value will have to earn a rate over time in excess of the risk free rate. Analysts tend to adopt as their measure the current rate on 'gilt edged' stock. These are generally held to be fairly liquid investments and are safe in money terms if held to redemption. Additionally, if they are sold at a loss, or at a gain, it can be reasonably assumed that there will have been a similar movement in the values of most other forms of investment. Property, being considerably less liquid and generally harder to sell, is expected to achieve a return over time of 1–2% above the going rate on gilt edged stock, or higher in the case of poor-quality properties or locations. This property risk premium is not a 'given' as under certain market conditions demand can be sufficiently strong to force prices up and yields down to the level of stock redemption yields or below. The valuer has to appreciate the full investment market not only to be able to value and reflect the changing market conditions but also to be able to advise investors on the appropriateness of investing in property in general and on specific individual property investment opportunities.

The DCF method requires the valuer to *'forecast'* rents and to discount those rents at a rate sufficiently higher than the risk free rate to account for the additional risks involved in the specific property investment. If there is no risk premium then the use of the DCF approach will make this fact explicit. 'Forecast' here does not mean prediction nor does it necessarily imply a projection based on extrapolating or extending the past into the future. It is an estimate of the most probable rent due in *n* years' time based on sound analysis of the past and present market conditions. The current preference seems to be to assess an implied growth rate for rent from the relationship between all risk capitalisation rates and gilts plus a risk premium of 1–2%.

Capitalisation and DCF approaches will be used in problems relating to the valuation of freehold interests in property. The examples will indicate the approach adopted.

Capitalisation and DCF methods

The technique of converting income into a capital sum is extremely simple. In the case of freehold interests in property the income will have the characteristics of an annuity which may be fixed, stepped, falling or variable.

A fixed or level annuity

If the income is fixed for a period much in excess of 60 years or in perpetuity, or if a property is let at its market rental and there is market evidence of capitalisation rates, then it can be treated as a level annuity in perpetuity.

Net income ÷ *i* = Capital value

Or Net income $\times \dfrac{1}{i}$ = CV

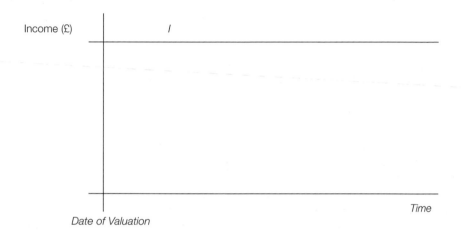

Example 6.1

A shopping centre was developed on a ground lease. The ground lease is for 150 years at a ground rent of £10,000 a year. The lease has no break clauses or rent review clauses. The lease has 110 years to run.

- Fixed Income of £10,000.
- Unexpired Term 110 years as at valuation date.
- Capitalisation rate 8%.

£10,000 ÷ 0.08 = £125,000

A conventional valuation would appear as:

Net Income	£10,000 a year
YP in perpetuity at 8% (Present value of £1 pa in perpetuity at 8%)	12.5
Estimated Market Value	£125,000

A perpetuity approach is used because the discount factor for 110 years is equivalent to that in perpetuity (see p11).

There may be market comparables to support the capitalisation rate of 8%. But in the absence of evidence the rate can be compared to that achievable as a return on irredeemable government stocks with an adjustment for any extra risks or security attaching to the property investment. Investments of this nature are very secure in money terms but not in real terms. The money security is provided by the fact that the ground rent is secured by the rents paid by the occupying tenants to the developer and by the fact that the ground lease would be forfeited in English Law on non-payment of the ground rent of £10,000.

A similar approach can be adopted for properties currently let at their market rental.

Example 6.2

Value a freehold shop in a prime trading position let at its market rental value of £20,000 a year on full repairing and insuring terms. The lease is for 20 years with reviews every five years. Similar properties are selling on the basis of an all risks yield (ARY) or capitalisation rate of 5%.

Rent	£20,000
YP perpetuity at 5%	20
CV	£400,000

Notes:
1. Where the lease is only for 20 years but it can be assumed that it would continue to let readily and that full rents will be receivable then it can be treated as a perpetual income. Such an assumption could not be made if the trading position was under threat from planned development such as an out-of-town centre.
2. Although the rent is reviewable every five years there is no need to provide for this explicitly in a capitalisation approach if at each reversion the rent would also be equivalent to today's market rental value of £20,000.
3. All property incurs a management cost but customarily when property is let on FRI terms no deduction is made for management.
4. For some purposes the costs of acquisition (survey fees, legal fees and Land Value Stamp tax) will be deducted to arrive at a net value. Where the net value plus fees adds up to £400,000, £20,000 will represent a return on total outlay of 5% (see Chapter 13 for more information on property investment analysis and advice). The return on net value would be greater than 5%.

Example 6.3

If the rent in example 6.2 is payable in advance and the market is still looking for a 5% investment the valuation becomes:

Rent	£20,000
YP perpetuity at 5% in advance	21
	£420,000

Notes:
1. £420,000 is being paid for the right to receive an immediate £20,000 and thereafter £20,000 at the end of each year. This could only be the case if the valuation was undertaken on the rent due date. In such a case it is not the terms of the lease but the date of the valuation and the rent payment date that guides the valuer. In property sales the contract will usually provide for rental apportionment as between vendor and buyer. The definition of market value implies contractual completion at the

date of valuation. The in advance rent would be paid to the vendor by the tenant and transferred to the buyer as part of the contract. Hence the extra £20,000 is paid to receive the immediate £20,000.

2. Similar points and changes can be made to reflect quarterly rental arrangements where the YP in perpetuity quarterly in advance would be used (see *Parry*'s 2003).

Valuers are being encouraged to represent their investment valuations as cash flows as this is a more familiar format for accountants and investment analysts.

Period	Net Income	PV £1 at 5%	NPV
1	20,000	0.95237	19,047.6
2	20,000	0.90703	18,140.6
3	20,000	0.86384	17,276.8
4	20,000	0.82270	16,454.0
5	420,000*	0.78353	329,082.6
			400,001.6

(* £420,000 represents the resale value at the end of year 5 being the right to continue to receive £20,000 pa thereafter in perpetuity plus the rent due at the end of year 5 of £20,000.)

This structure, set out on a spreadsheet, allows the valuer to revalue using a range of rates where there may be some degree of uncertainty as to the market rates (see Spreadsheet User 6 at the end of this chapter). The structure can also be lengthened and developed to reflect growth.

The use of non-conventional methods is inappropriate where there is ample evidence of recent transactions and where the capitalisation is of a property let at the current market rental value on current lease terms, however the following examples are included to indicate the development of the Short-Cut DCF/Equated Yield and Real Value Methods.

Short-cut DCF method

This method or approach was developed during the major property cycle boom years through the late 1970s and 1980s. The explanation that follows sets out a situation which was typical of that period when investors were buying for capital growth, prices were rising and yields had fallen far below the safe returns available from investing in fixed income securities such as gilt edge stock. The explanations and examples that follow have not been amended to reflect the very changed position of the 2000s.

In example 6.2 no account was taken of the fact that in year 5 there would be a review of the rent to the open market rental value. Fixed interest securities are considered to be risk-free; even more so where the value (stock face value) is index linked. So if at the time of valuation property of a particular type and quality is selling on a 5% basis and risk-free stock can be purchased to achieve a return of 10% then the market for property must be implying an expectation of growth. (The position in 2005, following the sharp decline in equity share prices, is very different in that interest rates have fallen, major investors have had to rebalance their portfolios, selling out of equities and buying in to stock, stock prices have risen with a commensurate decline in redemption yields and property market capitalisation rates have remained above gilt rates.) However when the reverse yield gap becomes pronounced as it did in the latter part of the 20th century the investor in property is buying for growth and this has to be interpreted carefully.

Thus £100 invested in perpetual stock at 10% will be producing £10 per year when the same invested in property is producing 5% or £5 per year. This represents a loss of £5 a year while the investor is accepting the greater risks attached to the purchase of property. If these greater risks warrant a 2% adjustment then the property investor should be looking for an overall return of 12% (10% + 2%). The difference between 5% and 12% is known as the *reverse yield gap*, being the opposite of the normal expectation that investors' returns rise as risks rise. The gap must be made good through growth.

The necessary level of growth can be calculated using DCF techniques, and a number of formulae and tables have been produced to simplify the process. In the above paragraph an annual growth rate of 7% might be inferred (12% – 5% = 7%) but in the case of property incomes the growth in rent can only be recovered at rent reviews which are rarely annual, more frequently five yearly but sometimes every three or seven years.

Given the rates of 12% and 5% the following growth rates would be implied given the following different Rent Review patterns:

Rent Reviews every :	Implied growth (12% and 5% pa)
3 years	7.32
5 years	7.64
7 years	7.93

Source: *Donaldson's Investment tables*

The same figures can be calculated from any one of a number of formulae:

$$K = e - (ASF \times P)$$

Where:

K	=	the capitalisation rate expressed as a decimal
e	=	the overall return or equated yield expressed as a decimal
ASF	=	the annual sinking fund to replace £1 at the overall or equated yield over the review period (*t*), and
P	=	the rental growth over the review period from which the implied rate of rental growth can be calculated.

Some commentators prefer the formulation:

$$(1 + g)^t = \frac{\text{YP perp at K} - \text{YP } t \text{ years at } e}{\text{YP perp at K} \times \text{PV } t \text{ years at } e}$$

Or

$$g = \left[\sqrt[t]{\left(1 + \left(1 - \frac{K}{e}\right)(1 + e) - 1\right)} \right] - 1$$

So given that K $=$ 0.05 (5%), e $=$ 0.12 (12%) and t $=$ 5 years
then:

$$0.05 = 0.12 - (0.1574097 \times P)$$
$$0.1574097\ P = 0.12 - 0.05$$
$$0.1574097\ P = 0.07$$
$$P = 0.07\ /\ 0.1574097$$
$$P = 0.444699\ \text{(which multiplied by 100 to convert to a \% = 44.47\% over 5 years)}$$

The nature of compound interest was outlined in Chapter 1. Here there is an implied growth in rent, P, over 5 years of 44.47% and therefore:

$$P = (1 + g)^{t} - 1$$

Where:

g $=$ growth per year expressed as a decimal and

t $=$ the rent review period (in this example this is every 5 years)

$1 + P$ must equal $(1 + g)^{t}$ to give

1.444699	$=$	$(1 + g)^{5}$
$1.444699^{0.20}$	$=$	$(1 + g)$
[(or $\sqrt[5]{1.444699}$	$=$	$(1 + g)$]
1.076355	$=$	$(1 + g)$
Therefore g	$=$	$1.076355 - 1$
$g\%$	$-$	7.6355%

The figure 7.6355% represents the implied annual average rental growth in perpetuity.

The steps in the Short-cut DCF method can now be stated:

- discount the current rent to the next review or lease renewal at the overall rate or equated yield. This is logical as there can be no growth in a rent fixed by legal contract (the lease) for a term of years
- multiply the Estimated Rental Value (RV) by the amount of £1 for the period to review at the implied rate of rental growth
- discount (capitalise) the future market rent in perpetuity at the capitalisation rate (All Risks Yield) and discount the capitalised value on reversion for the intervening period at the over-all or equated yield.

The issues to be resolved are:

- assessment of ARY (Capitalisation Rate)
- assessment of Overall Yield (Equated Yield).

Example 6.4

Revalue the property in Example 6.2 using a Short-cut DCF approach on the basis of an equated yield of 12% and an ARY of 5%.

		£20,000	
PV £1 pa for 5 years at 12% (YP)		3.6047	£72,094
		£20,000	
Amount of £1 in 5 years at 7.6355%		1.4447	
Implied Rent in 5 years		£28,894	
PV £1 pa in perp. at 5%	20		
PV £1 in 5 years at 12%	0.56743		
		11.3486	£327,906
			£400,000

The value is identical to the normal income capitalisation because of the underlying assumptions incorporated in the implied growth formula. This could be set out as a cash flow over a number of years if preferred by the client.

The figure of £28,894 does not represent a forecast of the rent to be expected in five years time. It is simply an expression of future rental derived from the market's implications of purchasing risk investments below risk free rates. It represents the rental needed in five years time to recoup the loss of return (income) during the next five years in order to achieve an overall return 2% higher than the given risk free rate.

Question 1:

Freehold shops in prime locations on FRI terms, with rent reviews every five years are selling on an initial yield (ARY) of 6%. If investors' target rates are 1% above gilts which stand at 8% what is the implied growth rate?

Real value method

Professor Neil Crosby (University of Reading) developed a real value approach based on earlier work of Dr Ernest Wood and has become the principal exponent of this technique. The technique is similar to the Short-cut DCF Method in that it values the current rent at the equated yield. Where it differs is that it retains the reversionary rent at today's market rental value but discounts the reversionary capital value at the Inflation Risk Free Yield (IRFY).

When using any of the capitalisation techniques the valuer needs to assess the nature of the rental agreement as set out in the lease to ascertain, in the light of current market conditions, the extent to which the rental income is inflation proof.

- Where rental income is completely inflation prone — that is where it is fixed with no expectation of change — the equated yield should be used in the discounting process.
- Where rental income is completely inflation proof the IRFY should be used.

This process reflects the fact that between rent reviews the real value of the income may be falling.

The steps in the method follow the DCF method in assessing the implied growth rate; from this the valuer calculates the IRFY from the formula:

$$\frac{e - g}{1 + g}$$

In this example

$$\text{IRFY} = \frac{0.12 - 0.076355}{1.076355} = \frac{0.043645}{1.076355} = 0.0405488 \times 100 = \textit{say } 4.055\%$$

the valuation becomes:

		£20,000	
PV £1 pa for 5 years at 12% (YP)		3.6047	
			£72,094
		£20,000	
PV £1 pa in perp. at 5%	20		
× PV £1 in 5 years at 4.055%	0.81975	16.395	£327,900
			£399,994

(small discrepancy of £6 due to rounding of numbers)

Both the DCF and Real Value models overcome some of the key criticisms of conventional valuation when valuing property currently producing less than their market rental value.

Conventional methods for valuing property investments with stepped incomes (generally known as reversionary property investments) are considered in the next section. As set out in the next section they can be criticised because they:

1. overvalue the fixed term income through the use of a capitalisation rate rather than the equated yield
2. undervalue the reversion by using market rents expressed in today's market terms (that is at the date of valuation) rather than reverting to future rents.

Point 1 is covered in both non-traditional methods by using the equated yield. Point 2 is covered in the DCF method by capitalising the assumed implied future rent using the capitalisation rate and by deferring using the equated yield; in the real value method by deferring the capitalised current rent at the IRFY.

Income patterns

A 'stepped' annuity

If the income is fixed by a lease contract for x years and is then due to rise, either by reversion to a rack rental to be valued, for simplicity, in perpetuity, or to rise to a higher level for y years, then reverting to a rack rental in perpetuity, the valuation may be treated as an immediate annuity plus a deferred annuity in one of two ways. The first is referred to as a term and reversion, the second as the layer method.

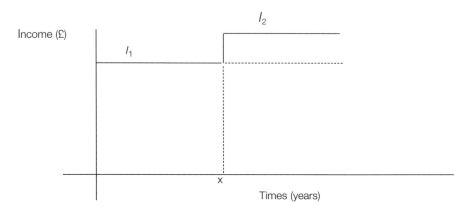

Stepped Annuity

Having estimated I_1 and I_2 as accurately as possible, the critical factor is the capitalisation rate $i\%$ which need not remain the same throughout. If it does remain the same, then each approach will produce the same value estimate.

Example 6.5

If the property in example 6.2 is let at £10,000 pa on FRI terms with two years to run to a review the valuation could appear as:

Term and reversion

Term	£10,000	
YP 2 years at 5%	1.86	
		£18,600
Reversion	£20,000	
YP perp at 5%		
× PV £1 in 2 years at 5%	18.14	£362,800
		£381,400

Layer		
Layer	£10,000	
YP perp at 5%	20	£200,000
Top slice	£10,000	
YP perp at 5% × PV £1		
in 2 years at 5%	18.14	£181,400
		£381,400

Notes:
1. Because all the capitalisation and deferment or discounting rates have been left at 5% both conventional methods produce the same value figure.
2. In this case YP x PV of 18.14 comes from the YP of 20 multiplied by the PV factor of 0.9070. It could have been found by using the YP of a reversion to a perpetuity table or deducting the YP for five years from the YP in perpetuity (see Chapter 1).

Here again a cash flow would represent this as:

End of Year	Net Income	PV £1 at 5%	NPV
1	10,000	0.95238	9,523.80
2	10,000	0.90703	9,070.30
3	420,000	0.86384	362,812.80
			£381,406.90

Some valuers still insist on using different rates to reflect some personal view on security of income. This can be dangerous and can produce very peculiar results. This can apply to either of the traditional approaches as illustrated below.

Term and Reversion		
Term	£10,000	
YP 2 years at 4%	1.89	£18,900
Reversion	£20,000	
YP perp def'd 2years at 5%	18.14	£362,800
		£381,700
Layer		
Layer	£10,000	
YP perp at 4%	25	£250,000
Top slice	£10,000	
YP perp def'd 2 years at 5%	18.14	£181,400
		£431,400

The variation in the term and reversion is negligible and in a valuation would not be material when expressed as an opinion of, say, £380,000. In the layer method, failure to appreciate the effect of the change results in an underlet property being valued higher than the equivalent fully let property (see example 6.2). This cannot be correct. Practising valuers generally avoid such an error by capitalising the layer income at the ARY and increasing the rate on the top slice.

The reason for valuers wishing to change rates within a valuation rests with the historic evolution of the methods and the changing economy as reflected in the investment market (see Baum and Crosby, 1995).

Valuers in the 19th century were primarily dealing with rack rented properties, largely agricultural, at a time when inflation was relatively unknown. A realistic and direct comparison could be made between alternative investments such as the undated securities (gilts) issued by government, the interest on deposits and farmland. The latter, being more risky, was valued at a percentage point or so above the secure government stock. Other property, such as residential and the emerging retail properties, were more risky but were let on long leases so they could be capitalised in perpetuity, again at an appropriate higher rate.

In the 1930s the depression made any contracted rent a better or more secure investment than an unlet property and there was real fear that at the end of a lease the tenant would seek to redress the position by requesting a lower rent or by vacating. Logic suggested the use of a lower capitalisation rate for secure contracted rents than unsecured reversions.

This approach continued unchallenged until the 1960s. In the meantime the market had changed in many ways and with the rise of inflation the investment market had identified a crucial difference between fixed-income investments such as gilts and those where the owner could participate in rental and capital growth such as equities and rack rented properties.

This change was noted in the market and the reverse yield gap emerged. Nevertheless, valuers continued to use a capitalisation approach, varying yields to reflect the so-called security of income of contracted rents, the argument being that a tenant would be more likely to continue to pay a rent which was less than the market rental value because it represented a leasehold capital value (see Chapter 7). The market did not appear to recognise that the technique of the 1930s was one of using rates to reflect money risk, whereas those after the appearance of the reverse yield were capitalisation rates which reflected money risks and expectations of growth.

There are currently many arguments for not varying capitalisation rates in simple conventional valuations, but two will suffice. By the 1960s the property market for income-producing properties was being dominated by the major institutions.

Property investment surveyors were increasingly required to specify in their reports:

(a)   the initial or year one rate of return and
(b)   the investment's internal rate of return (IRR).

The former acted as a cut-off rate as actuarial advice at that time required all investments, depending upon the fund, to produce a minimum return — frequently 3–4%. It still acts as a cut off rate as some investors will not purchase unless the initial yield is above an actuarially specified minimum. The second caused confusion for valuers used to using perhaps four or five rates in a multiple reversionary property. The confusion arose because few valuers understood the concept of IRR, few knew how to calculate it, and in the 1960s there was nothing more sophisticated than a slide rule to help with the calculations.The profession took a little longer to recognise that if the capitalisation rate used in the valuation was held constant it would be the investment's expected IRR.

It can also be shown that in most term and reversion exercises where variable rates are used the IRR — now popularly called the equivalent yield — is almost the same as the ARY. As a result many valuers, whether using the term and reversion or layer method, now use an equivalent or same yield approach. It is, however, important to distinguish between the terms 'equivalent' yield and 'equated yield' .

In *Donaldson's Investment Tables* it is stated that:

The equated yield of an investment may be defined as the discount rate which needs to be applied to the projected income so that the summation of all the income flows discounted at this rate, equals the capital outlay; ... whereas an equated yield takes into account an assumed growth rate in the future annual income, an equivalent yield merely expresses the weighted average yield to be expected from the investment in terms of current income and rental value, without allowing for any growth in value over the period to reversions.

Secondly it can be shown that variable rates pose their own problems.

Bowcock (1983) uses a simple example of two properties each let at £100 a year but one with a review in five years and the other with a review in 10 years, both with a reversion to £105 a year.

Traditionally the solution might appear as:

Term	£100	
YP 5 years at 9%	3.8897	
		£388.97
Reversion	£105	
YP perp def'd 5 years at 10%	6.2092	£651.96
		£1040.93
Term	£100	
YP 10 years at 9%	6.4177	£641.77
Reversion	£105	
YP perp def'd		
10 years at 10%	8.8554	£404.82
		£1046.59

Clearly, there is something wrong with a method which places a higher value on the latter investment, which includes a ten-year deferred income, than the former with a five-year deferred income. In practice, this potential error is not normally noted by valuers, who subjectively adjust their rates to reflect (a) their view of the extent of the security of income, namely the difference between contracted rent and market rental value, and (b) the risk associated with the period of waiting up to the rent review.

Some critics have also drawn attention to the problems of selecting the correct deferment rate in a variable yield valuation. Customarily the reversion has been deferred at the reversionary capitalisation rate. If this is done then it can be shown that the sinking fund provision within the YP single rate formula (at 9% in the above example) is not matched by the discounting factor (10% in the example); as a result the input rates of return of 9% and 10% will not be achieved. This can be corrected by deferring the reversion at the term rate.

The discerning valuer must conclude that it is safer and more logical to adhere to the same yield or equivalent yield approach. It is also the easiest yield to extract from sales evidence as the calculation is identical to that of the IRR. On a same yield basis there is no distinction to be made between the two capitalisation methods. However, the term and reversion seems the most acceptable theoretical method and preferable for valuations involving lease renewals when a void allowance or refurbishment cost may need to be built in. It is also easier to handle if outgoings have to be deducted. The layer method is possibly simpler and more useful when handling certain investments where clients wish to know what price has been or is to be paid for the top slice an example would be the valuation of over rented property where the top slice might cease at the lease renewal when the rent reverts to the lower market rent.

Valuers are still faced with the problem of finding comparables. The more unusual the patterns of income the more difficult it is for the valuer to judge the correct capitalisation rate.

For example, what would be the right equivalent yield to use for a property let at £10,000 a year with a reversion to £20,000 a year in 15 years?

The problem in these cases is that the period to the reversioin is beyond the normal rent review or lease length without any opportunity for the rent to be adjusted for inflation or growth. The investment is more risky in money terms it is not investment proof. In these cases an equivalent yield approach is likely to be criticised if the yield used is taken direct from the typical market and not from a comparison with similar long reversionary properties. The market solution is to intuitively adjust the capitalisation rate upwards. The question the valuer has to decide is how big should the upward adjustment be, and does it increase with the length of the reversion or with the scale of the reversion or with both. For example should a different view be taken of a reversion in 15 years where the increase in rent will be, in today's terms, 100% to one where the increase is 500%.

The position during periods when there is a positive reverse yield gap as there was in the 1980s/1990s was one where the DCF approach or real value approach could be used to advantage. This is still the case where investments with growth potential are selling on a low yield basis compared to the fixed income returns on government stock. In such cases the risk position has not changed, as all property is still riskier than holding cash or gilt edge stock. Thus if gilts are at say 12%, as they once were, a property investor should arguably be looking for 13–14%. Purchasing at 5% is forgoing at least £8 per £100 invested each year. Few people offered employment at £13,000 a year would reject it in favour of one at £5,000 a year unless they were certain that over a specified contract term the loss of £8,000 a year would be compensated by regular and substantial increases in salary. The same is true of property. So capitalising at 5% when redemption yields on gilts are substantially higher must imply rental growth in perpetuity and the first criticism of the equivalent, or same yield approach for reversionary investments, is that it fails to indicate rental growth explicitly by using a reversion to current market rents.

Also, both the term income and the reversion are capitalised at the same ARY implying the same rental growth when clearly a lease rent is fixed by contract for the term. The only substantive arguments in defence of the equivalent yield approach are simplicity and that purely by chance it may overvalue the term just sufficiently to compensate for any undervaluation of the reversion (see Baum and Crosby, 1995). The modified or short cut DCF and Real Value approach have been developed to overcome these criticisms.

Example 6.6

Using the facts from example 6.5, revalue on a modified DCF basis assuming the MRV of £20,000 is based on a normal five-year review pattern and that an equated return of 12% is required

The implied growth rate is 7.6355% (see p97)

This implies that the rent in 2 years time would need to be:

£20,000 × amount of £1 at 7.6355% for 2 years
£20,000 × 1.5854 =£23,170.80, say £23,170

The property can now be re-valued on a contemporary basis, in this case using a modified or short-cut DCF or Real Value method (Baum and Crosby, 1995, provide a full critical comparison).

Short-cut DCF

First 2 years	£10,000	
YP 2 years at 12%	1.6901	
		£16,901
Reversion to	£23,170	
(£20,000 × A £1		
at 7.6355% for 2 years)		
YP perp at 5%	20	
	£463,400	
PV £1 at 12% for 2 years	0.7972	£369,422
		£386,323

The difference in opinion in practice on such a short reversionary property would be less significant as the valuations would probably be rounded to £382,000 (example 6.5) and £385,000 respectively. However, it can be seen how the term and reversion produces a nearly acceptable solution by overvaluing the term by £1,699 (£18,600 − £16,901) which to some extent compensates for the undervaluation of the reversion by £6,622 (£369,422 − £362,800).

The short-cut DCF now reads like an investment valuation as all contracted income is discounted at a money market rate or target rate(the valuer's equated yield) and the reversion is to an expected, albeit implied, rent, not today's rental value.

The importance of this issue is more evident when considering a longer reversionary property.

Example 6.7

Assuming the same facts as examples 6.5 and 6.6 but assuming a reversion in 15 years.

Conventional
Term and reversion

	£10,000	
YP 15 years at 7%	9.1079	
		£91,079
	£20,000	
YP perp at 7%	14.2857	
	£285,714	
PV £1 in 15 years at 7%	0.3624	£103,542
		£194,621

A capitalisation rate of 7% has been used rather than 5% in order to reflect the difference between an investment with a long reversion and one with a short reversion.

However:
£20,000 × Amount of £1 at 7.6355% for 15 years
£20,000 × 3.02 = £60,306.34

Short-cut DCF

	£10,000	
YP 15 years at 12%	6.8109	£68,109
	£60,306	
YP perp at 5%	20	
	£1,206,120	
PV £1 in 15 years at 12%	0.1827	£220,358
		£288,467

Here a subjective assessment had to be made as to the appropriate equivalent yield to use for a 15-year deferment. The modified DCF assesses the fixed lease income at the equated yield or opportunity cost, the reversion is to the implied rent in 15 years time and the value is deferred at the equated yield. If the DCF is arithmetically more correct (and it is) then an equivalent yield of less than 5.25% would have to be used to arrive at a figure of £288,467. It is extremely difficult in the absence of true comparables to arrive at a correct equivalent yield subjectively. Intuition appears to be the favoured approach of a number of valuers with some using money market rates for discounting the fixed term in an effort to overcome the weaknesses of the conventional valuation method.

Currently, though, it is not possible to say that a short-cut DCF approach must be the preferred approach. It may be the more rational approach in that it is possible to argue that investors should be indifferent between the short and long reversionary properties if they are expected to produce the same equated yield. But this is difficult to support if the market is substantially discounting long reversions through the subjective approach of their investment surveyors; implying inconsistency over the choice of equated yields. The short-cut DCF can be proved by using a full projected cash flow over, say, 100 years, discounted at 12% as in Table 6.1.

In practice such contemporary methods seem to be rejected in favour of the market methods. This implies that although valuers do not make the market they do in some instances have a strong influence. Current market practice favours the simple equivalent or same yield approach.

This creates the probability that investors will conclude that some valuers are overvaluing, or undervaluing, properties and will sell or purchase in the market at market prices to take advantage of such market imperfections.

Throughout this section it must be remembered that the growth rates used are implied and that the figures derived in no way predict the future. They merely provide the valuer with an additional tool with which to examine and, as will be seen later, to analyse the market. Such implied rents must be critically examined against the reality of the marketplace and realistic economic projections.

Table 6.1

DCF to 100 years allowing for a rental growth at 7.6355% discounted at an equated yield of 12% and providing for a reversion to future capital value in year 100. This would bring the value figure more into line with the modified DCF.

Period (years)	Income × A£1 at 7.6355%	PV £1 pa at 12%	PV £1 at 12%	Present value
0–15	10,000	6.8109	–	68,109
16–20	60,306	3.60477	0.18269	39,714
21–25	87,124	3.60477	0.10367	32,558
26–30	125,868	3.60477	0.05882	26,688
31–35	181,842	3.60477	0.03337	21,874
36–40	262,706	3.60477	0.01894	17,936
41–45	379,532	3.60477	0.01074	14,693
46–50	548,310	3.60477	0.00609	12,037
51–55	792,144	3.60477	0.00346	9,880
56–60	1,144,410	3.60477	0.00196	8,085
61–65	1,653,329	3.60477	0.00111	6,615
66–70	2,388,566	3.60477	0.00063	5,424
71–75	3,450,761	3.60477	0.00036	4,478
76–80	4,985,314	3.60477	0.00020	3,594
81–85	7,202,282	3.60477	0.00011	2,855
86–90	10,405,137	3.60477	0.00006	2,250
91–95	15,032,301	3.60477	0.000037	2,005
96–100	21,717,164	3.60477	0.000021	1,644
100	434,343,280	3.60477	0.00001	4,343
				Value £284,782

Example 6.8

Value freehold shop premises let on lease with four years to run at £7,000 a year. The tenant pays the rates and the insurance and undertakes internal repairs. It is worth £12,500 a year net today and rental values for this type of property are continuing to rise in this area.

Assumptions:

1   Current capitalisation rates on rack rented comparable premises are 8%.
2   Rents are due annually in arrears.

Market Valuation:

Gross income per year for four years		£7,000
Less landlord's outgoings		
External repairs and decorations[1]	£650	
Management at 5% of £7,000[2]	£350	£1,000

Net income	£6,000	
Years' purchase for 4 years at 7%[3]	3.39	£20,340
Plus reversion in 4 years to[4]		
Net income	£12,500	
YP in perp def'd 4 years at 8%[5]	9.19	
		£144,875
Estimate of capital, market value		£135,215

Value (say) £135,000[6]

Notes:
1   Based on office records, etc.
2   Based on fees charged by management department for comparable properties. These are normally based on the service provided rather than the percentage of rent collected.
3   The relationship between the rates here and later (7% and 8%)depends upon a number of assumptions as to the true behaviour of the market. The argument here is that the £6,000 is more secure in money terms, hence a lower rate is used.
4. In a normal market valuation the reversion is to rental value (MRV) as estimated in today's terms. There is no deduction here as it is a net rent, although some valuers could allow for management at say 1–2% of rent collected.
5   YP in perp deferred = YP perp. at 8% × PV £1 in four years at 8% or YP perp. at 8% less YP for four years at 8%.
6   This final estimate is rounded as no valuer can truly value to the nearest £; figures of this magnitude would bear rounding to the nearest £1,000.

The idea of reducing the term capitalisation rate by 1% to reflect money security is almost certainly inappropriate during inflationary periods. Money security should give way to a reflection of purchasing power risk. Some purchasers will not differentiate between a property let below market rental and one let at market rental, because there is no real difference provided the sum paid for the investment reflects the difference in current income and any difference in expectation of future rental change. But, on the latter point, is it reasonable to use the same discount rate for both the capitalisation of, and the four-year deferment of, the reversionary income?

A number of valuers would comment that this rate and the deferment rate should be higher by 1-2% to reflect the greater uncertainty of receiving the future increased income. Of the two, the concept of increasing the deferment rate may seem the most logical as 8% could be applied to £12,500 if it was receivable today. But those who defend the former method argue that it results in an overvaluation of the term which helps to compensate for the undervaluation of the reversion.

The point here in practice, and as indicated in the question, is that if the rent of £12,500 is continuing to rise the true rent in four years time will exceed £12,500 and the reversion has therefore been undervalued. Those who favour a DCF approach would argue that neither approach is defensible because, assuming the opportunity cost of money is 12%, capitalisation rates of 7% and 8% and deferment rates of 8% imply a particular expectation of growth in the incomes over the periods involved.

The particular rates used in the examples may fortuitously have produced a result which is close enough to the correct value of the property.

A short-cut DCF approach may be preferred or used as a check.

A comparable conventional layer method valuation employing different rates may be laid out as in example 6.9.

Example 6.9

Layer income	£6,000	
YP perp at 7%	14.29	
		£85,740
Marginal income[1]	£6,500	
YP perp def'd 4 years at 8%	9.19	
		£59,735
Estimate of capital/market value		£145,205

Value (say) £145,000

Note:
1 Reversionary income of £12,500 minus initial (layer) income of £6,000.

An equivalent yield valuation may be derived from either approach to find the single equivalent rate of interest which, if applied to both the term and reversion (or layer and marginal) income, will produce the same valuation.

Example 6.10

Valuation		£135,215
Term:		
Net income	£6,000	
YP 4 years at x%	a	£ A
Reversion:		
Net income	£12,500	
YP perp def'd 4 years at x%	b	£ B
	£(A+B) = £135,215	

By trial and error, x% may be found. In this case it is approximately 7.97%. This is the equivalent yield.

Valuation (equivalent yield)
Term:
Net income	£6,000	
YP 4 years at 7.97%	3.31	£19,860
Reversion:		
Net income	£12,500	
YP perp def'd 4 years at 7.97%	9.23	£115,375
		£135,235

Valuing throughout at 8% produces:

£6,000 × 3.3121 + £12,500 × 9.18787 + £19,872 + £114,848 = £134,720 or say £135,000

This suggests that making a 1% allowance from the ARY for security of income is of minimal consequence and as such of little purpose.

An equated yield valuation is a valuation which employs DCF techniques. This may be combined with a conventional valuation approach as a short cut or carried out fully as previously illustrated.

The conventional capitalisation approach is perfectly acceptable under normal conditions where property is let on normal terms with regular rent reviews, and where there is evidence of capitalisation rates. In such cases it is most logical to use an equivalent yield approach, whether it be in the format of the term and reversion, or the layer, method.

Where the income pattern is not normal, there is a strong case for using the DCF approach. DCF directs the valuer to concentrate upon an explicit consideration of the net current and future incomes, and upon the correct rate of interest to use to discount that specified cash flow.

A falling annuity

This can be treated in a similar manner to a rising annuity using either of the previous methods.

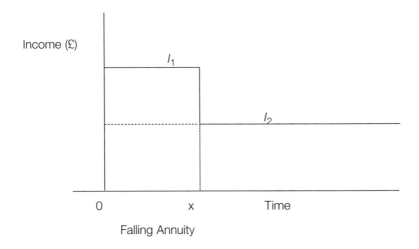

Falling Annuity

A number of valuers would adopt the layer method so that a higher capitalisation rate could be used for the income that will cease $(I_1 - I_2)$ as it is considered to be at greater risk.

The issue of falling incomes is not new as it tends to occur when economic slump follows economic boom.

Example 6.11

Value a freehold office building let at £10,000 a year. with five years to the review. The current market rental value is £7,500 a year. (The rental pattern is illustrated below in figure 6.4.)

The valuer's immediate response is 'will rents rise to the contracted level by the review date?' If he/she feels confident in saying yes, then the valuation problem is partly removed and £10,000 pa could be treated as a perpetual income provided a current capitalisation rate can be derived from market comparables.

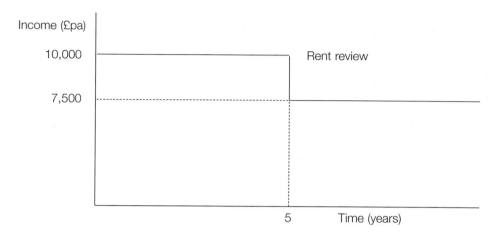

Rental pattern for example 6.11

The second question is whether the tenant can insist on a lower rent at the review date. If there is an upwards-only rent review clause then this might preclude a fall in rent.

The valuer's conventional solution to a terminating income is to use the layer method capitalising the income to be lost at a higher capitalisation rate.

Continuing income	£7,500	
YP perp at 8%	12.5	£93,750
Plus income to be lost	£2,500	
YP for 5 years at 10%	3.7908	£9,477
		£103,227

The £2,500 has been capitalised at a higher single rate, but there is a danger of double counting as the use of 8% during the term could be implying that the figure of £7,500 is growing and so a part of the term income has been valued twice. (A dual rate approach could be used for the £2,500 for five years but is not recommended.)

The valuer needs to distinguish the genuine terminating income from that which may represent a temporary shortfall. Thus a building with a time limited planning consent may display an income pattern which is going to fall and this reduction will not be recovered until the lower valued use has, through real and inflationary growth over several years, risen to the higher value use. In these cases a hybrid capitalisation may be necessary to reflect not only the fall in income but also the variations in 'risk' attaching to the tranches of income arising from different uses.

The 1979 edition of this book suggested that a DCF approach would be the preferred solution to the falling income problem. Crosby, Goodchild and others have developed this approach to the valuation of the over-rented property.

The steps in the solution are:

1. assess the implied rental growth rate from comparable investment properties let at their open market rental value
2. calculate the point in time when current open market rental value will exceed the current contracted income
3. capitalise the contracted rent up to the cross-over point or first review thereafter at the equated yield
4. capitalise the implied market rent at the equivalent yield (ARY) from the date taken in step 3 in perpetuity discounted for the period of waiting at the equated yield.

Solution to example 6.11 using the Crosby approach :

1. ARY from comparable properties are 8%
2. equated yield is 12%
3. implied rate of rental growth given $e = 0.12$ and $k = 0.08$ is 4.63%
4. 7500 × Amount of £1 in n years at 4.63% = £10,000

7500 × 1.334	=	£10,000
$(1 + 0.0463)^n$	=	1.334
n	=	$\sqrt[n]{1.334}$
n	=	>6<7
say 6.5 years		

5. In this example, given an upward only rent review in five years, the rent will remain at £10,000 at the first review. The valuer must exercise judgement to decide the most probable date for a review to occur. In this case the valuation might be:

Rent		£10,000	
YP 7 years at 12%		4.5638	£45,638
Rent		£10,000	
YP perp at 8%	12.5		
PV £1 in 7 years at 12%	0.45235	5.6543	£56,543
			£102,181

If the lease was due to terminate after five years then the implied rent at that time could be calculated at the implied growth rate of 4.63%. Then:

£7,500 × amount of £1 in 5 years at 4.63% = Implied rent in year 5.

The contracted rent would be valued at the equated yield for five years with a reversion to the lower rent in year 5. Here there are practical difficulties for the valuer as most landlords would wish to retain the higher rent by payment of a reverse premium in order to underpin future rent reviews, but valuers must always be conservative in their assumptions in a market valuation.

While theoretical solutions to this problem can be found using implied growth rates and DCF techniques these may not truly reflect behaviour in the marketplace and the valuer must not let the mathematics override market judgements. In particular future rents calculated using implied growth rates over the short term must not be used as a substitute for market analysis. The latter may well suggest minimal change in rental value over the short term.

The problem of falling incomes becomes more acute in the secondary and tertiary markets; that is, when considering over-rented properties in non-prime locations, or with non-prime tenants. Here market capitalisation rates will generally be much higher than gilt rates and the valuer will have to use his or her best judgement in circumstances where the calculation of implied growth is unrealistic and impracticable.

Question 2 Value a freehold warehouse

A comparable property on the same business park let at its open market rent of £42,500 has just been sold for £472,000. The subject property is similar in size but was let on a reduced rent of £30,000 three years ago. There is a rent review in two years time. Assess the ARY and value the warehouse. Demonstrate the use of modified/short cut DCF and Real Value approaches on the basis of an equated yield requirement of 12%.

Baum and Crosby (1995) suggest that

> for a property over-let to a very secure covenant under a lease with more than 15 years unexpired (on upwards only rent reviews), the term yield could be determined in comparison with a fixed income gilt plus a small risk margin for property illiquidity but without the traditional risks of property such as uncertainty for future income flow.

This is an interesting 1990s' observation which flows logically from earlier observation in this book that

> valuers in the nineteenth century were primarily dealing with rack rented properties, largely agricultural at a time when inflation was relatively unknown. A realistic and direct comparison could be made between alternative investments such as undated securities (gilts) issued by government. Property being more risky was valued at a percentage point or so above the secure government stock.

The market in the early 2000s again suggests that good quality property is selling at or just above the gilt rate whether or not it is over rented provided the lease term is reasonably long and the tenant is secure.

In some respects history is repeating itself. It is again reasonable to argue that the original ' term and reversion' valuation may be appropriate in some circumstances. Thus an over-rented property may show little prospect of income change to the owner until a future lease renewal at which point best estimates or implied growth may suggest a market rental above the current contracted rent. The future rent may appear to be less secure as it will depend on a new letting or lease renewal and at such a date in the future that further rental growth may be unlikely due to depreciation and economic obsolescence. A market conventional approach might be to capitalise the term at say 1% above comparable gilts reflecting current income security of the tenant in occupation, and the reversion at 2% above gilts to reflect the greater risk attached to the reversion including the risk of a letting void.

The reader is reminded that, while detailing income valuation methods and the links between conventional and DCF techniques, the real task is the identification of the risks associated with the property to be valued, which requires research into future expectations (forecasting) as well as observation of the current market and market transactions.

A variable annuity

It is fairly rare to find a completely variable income from property, ie one where the income changes from year to year. But the suggested technique is to treat the calculation as a Present Value calculation treating each payment as a separate reversion and discounting each to its present value at an appropriate rate of discount *i*.

Advance or arrears

It has always been the custom to assume annual 'in arrears' income flows and annual rates for capitalisation purposes. There is, however, a growing tendency, as property is let on 'in advance' terms, to value on such a basis. In this respect it is essential to determine the valuation date at the commencement, as in many cases this will fall between rental payments. Technically this is still an 'in advance' valuation as the assumption of open market value will generally imply an apportionment of rents received 'in advance' as at the date of completion of a sale, and for valuation purposes the valuation date may be treated as a sale completion date.

Second, it is necessary to be certain that in this respect the lease terms are being enforced. It is common to find rent due on 1 January 'in advance' being paid at a date later than 1 January.

Third, one should not lose sight of the fact that 'in advance' means one payment due immediately, ie its present value is the sum due, and thereafter the same sum is due for *n* periods in arrears.

The crucial point is the relationship between the income and the discount rate. The latter must be the correct 'effective' rate for the particular income pattern. For preference this should be derived from market analysis of comparable 'quarterly', 'yearly', 'in advance' or 'in arrears' transactions as investors may not in fact be prepared to accept the same effective rate.

It should be noted that the use of 'quarterly in advance' valuation tables for the valuation of incomes received on this basis may be questioned, as it is very rare that the valuation date will coincide exactly with the date when an instalment of rent is due. The income pattern may in fact resemble a 'quarterly in arrears' income.

If a property valuation is analysed on a precise basis, allowing both for the correct apportionment of rent as at the date of valuation and for the correct timing of future rents, then the rate of return will be higher than the rate per cent adopted in the valuation on an 'annual in arrears' basis. This realisation has encouraged some valuers to switch to 'quarterly in advance' valuations using published tables or programmed calculators and computers. Where this is done, valuers must be sure that they adopt the proper market relationship between income patterns and yields of comparables in their valuation work.

When the whole market relates to 'quarterly in advance' lettings and all valuers are analysing yields on a precise basis, then the market yields adopted by valuers for capitalisation work will be correct for 'in advance' valuations. Until then the valuer needs to be fully aware of the basis of the quoted market yield before transposing it to an 'in advance' valuation.

The reader should note that as the assessment of rental value, outgoings and capitalisation rates are all opinions, switching to 'in advance' valuation tables will not in itself achieve a better opinion of value. Attempts at such arithmetic accuracy may be spurious.

Save where stated, the 'annual in arrears' assumption has been used in all examples. This assumption is still commonplace in valuation practice, but accurate investment advice requires that estimation of the exact timing of income receipts is necessary in order to assess yields accurately. The use of spreadsheets enables the valuer to incorporate an accurate calendar within any valuation or analysis programme.

On examination it will be noted that certain property investments resemble certain other forms of investment and they can be distinguished by the future pattern of returns. Thus, property let on long lease without review could be compared to an irredeemable stock.

Owner-occupied freehold commercial properties are comparable to equity shares, whereas freehold properties let on a short lease or with regular rent reviews (while in a sense resembling equity investments) must also reflect the stepped income pattern. Short fixed-income leasehold interests are comparable to any fixed-term investment such as an annuity.

Recognition of these relationships is essential if the valuer is to make correct adjustments to the capitalisation rates to be used in a valuation where the income pattern produced by the property is out of line with current market evidence.

A final adjustment

The last steps in any freehold valuation are:

- to reflect on the final figure and review the value sum in the light of market knowledge, some valuers might call this the feel right factor
- to adjust the final figure to indicate that valuation is not a precise science and any value given to the nearest £ would suggest a high degree of accuracy, adjustment could be to round to the nearest appropriate round number given the magnitude of the valuation eg if the calculation comes to £10,565,212 the valuer might round to £10,550,000. There is no rule on this and only practice and experience can guide the valuer
- to adjust the figure before or after rounding to reflect that any acquisition of property will incur considerable on costs to cover Stamp Duty Land Tax, solicitors fees and associated charges, building surveyor and valuation surveyor fees; some of which attract VAT at 17.5%. SDLT is at 4% on sales above £500,000 and an overall adjustment of 5.75% is made on valuations at this level or of 4.75% if the valuation is between £250,000 and £500,000. The need for this adjustment is because a sale at say £1,000,000 for a income of £100,000 on a 10% capitalisation basis becomes a purchase at a total cost of £1,057,500 which provides a return of 9.46%, which is not what the valuation implied. A division of the figure by 1.0575 produces the sum which if it becomes the contract price will provide the investor with a 10% return on total cost. £1,000,000/1.0575 = £945,626 and £945,626 + 5.75% = £1,000,000.

These last steps are typical but given that valuation is both an art and a science some valuers will go no further than refining their opinion based on experience of the market place — the feel right factor.

Sub-markets

The RICS requires its members to accept instructions to value only if they have the expertise within the specified market to be able to provide professional advice. In this respect it is worth noting, that while the general principles of freehold investment valuation methodology will apply to all parts of the market, that expertise or market experience is often very specialist and may relate to a relatively small sub-markets or sector.

The following chart is indicative of some of the larger sub-markets.

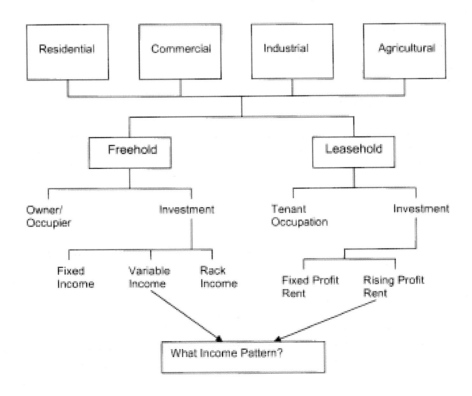

The valuation of leasehold interests is considered in the following chapter.

Summary

- All freehold income valuations are based on the principal of discounting future net benefits to their present worth.
- Two broad methods are used:
  - conventional capitalisation at the All Risks Yield or Equivalent Yield
  - contemporary DCF/Equated yield or Real Value approaches
  - both can be expressed in Cash Flow Format.
- Conventional methods are used where there is good evidence of capitalisation rates from the market.

- Contemporary methods are used where there is good evidence of capitalisation rates but where it is desirable to express rental changes explicitly and
  - for the assessment of worth to a client given the client's target rate
  - where there is poor market evidence, for example, in the case of long reversions and over-rented properties.
- Assessment of implied growth rates for use in contemporary methods relies in part on market evidence of capitalisation rate and on the valuer's ability to interpret the market's requirements in terms of equated yields. Baum has demonstrated that variations in the latter are less significant than expected for most valuations.
- Assessment of implied growth rates must not be confused with forecasts of future rents based oneconometric projections.

Spreadsheet User 6

To assist in the analysis of property sales and in the valuation of a typical property investment the valuer may wish to develop simple Excel spreadsheets.

The following spreadsheet shown below can be used to assess equivalent yields on a trial and error basis if a purchase price is known. Alternatively, if the valuers task is to assess market value then the structure of this spreadsheet will allow for this function to be performed and for market values to be calculated.

Project 1

The spreadsheet reprinted here is an illustration of the learning tools that readers, in particular students, can develop for themselves to facilitate the calculations required to complete typical valuation tasks. This spreadsheet is designed to calculate an equivalent yield by trial and error and to perform a term and reversion valuation on a same or equivalent yield basis.

The first section contains the data required for both calculations. Cells are completed for the data headings and left blank for the subject property data details. Typical costs have been inserted but all of these can be changed given the individual market circumstances. The calculations that follow the data, access the cell data as part of the equations set up in each blank cell.

Term and reversion

All data boxes need to be completed other than the contract price cell. Scrolling down the spreadsheet transfers the data to the respective cells and the equations created in the YP cells produce the YPs for the term at the Trial Rate (Equivalent Yield) and for the YP in perpetuity deferred. The value of the term and the reversion are then computed and summed in the cell against market value.

It is regular practice to reduce the market value figure for the typical acquisition costs. By so doing the valuer retains the rationale of the valuation in that acquiring at the netted down figure then incurs acquisition costs, when these are added together the result equals the already determined Market Value and hence the equivalent yield is as specified in the trial rate cell. If this is not done then once the costs are added to the Market Value the non growth internal rate of return of the investment will be less than the trial rate.

Equivalent Yield analysis

If a freehold sale price is known together with the current net of outgoings rent, the length of the unexpired term or period to rent review and the market rental value then the spreadsheet can be used iteratively on a trial and error basis to find the equivalent yield.

The net present value cell is the Market Value at the trial rate less the Contract Price (sale price of the property). If the NPV at the trial rate is zero then the trial rate will be the equivalent yield. If the NPV is positive the trial rate is too low, if negative then the trial rate is too high. By changing the trial rate and setting the degree of accuracy (number of decimal places for the rate per cent) the NPV can be quickly and repetitively calculated until a Zero NPV or near Zero NPV is found.

Other yields

The final cells present the simple assessment of initial yield and Yield on Reversion which are often quoted by investment agents in sales details and normally requested by investors given the details as to purchase price.

Readers who create a similar spreadsheet can experiment with colour, layout and other features of Excel.

Project 2

Readers who are studying the subject of Property Investment Valuation should develop spreadsheets as in Project 1 and then extend these to cover:

- term and Reversion on a quarterly in advance basis
- similar spreadsheets using the layer method
- spreadsheets which develop the MRV in the data section from areas and rents per m^2
- spreadsheets which allow for rent on internal repairing terms with provision for assessing the net of outgoings income on screen
- implied rental growth spread sheets
- spreadsheets to undertake valuations using Modified DCF or other course preferred non conventional methods.

A spreadsheet for Market Rented valuations on a perpetuity basis is probably going to be the simplest starting point in creating a personal set of spreadsheets.

Whenever a spreadsheet is used the user has to be satisfied that figures produced are correct. A spreadsheet will produce correct mathematical solutions given the equations the creator has built into the structure which may not be what is required and may not be a correct 'opinion of Market Value'.

Spreadsheets can be developed for Leasehold Valuations but these will depend upon the valuer's preferred methodology. That methodology debate is set out in Chapter 7 and the authors leave spreadsheet development in that area entirely with the readers imagination.

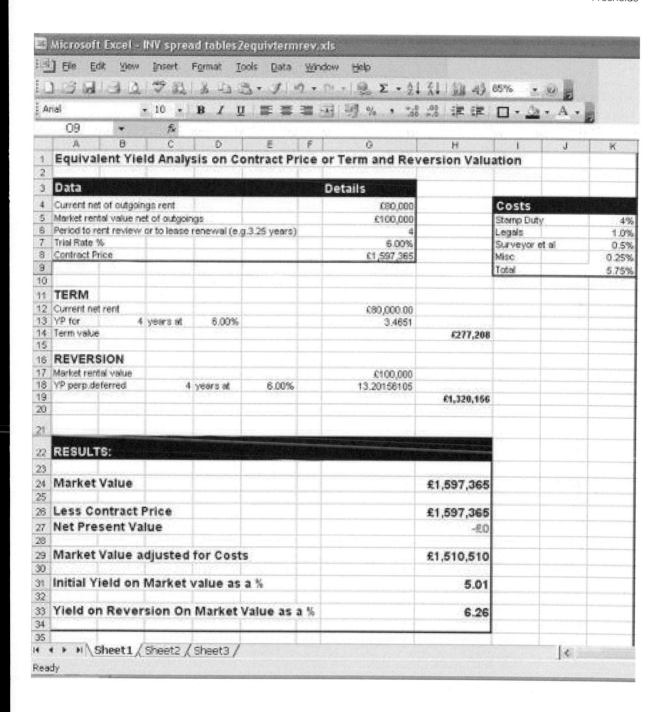

Microsoft Excel - INV spread tables2equivtermrev.xls

File Edit View Insert Format Tools Data Window Help

Arial 10 B I U %

O9

	A	B	C	D	E	F	G	H	I	J	K
1	**Equivalent Yield Analysis on Contract Price or Term and Reversion Valuation**										
2											
3	**Data**						**Details**				
4	Current net of outgoings rent						£80,000		**Costs**		
5	Market rental value net of outgoings						£100,000		Stamp Duty		4%
6	Period to rent review or to lease renewal (e.g.3.25 years)						4		Legals		1.0%
7	Trial Rate %						6.00%		Surveyor et al		0.5%
8	Contract Price						£1,597,365		Misc		0.25%
9									Total		5.75%
10											
11	**TERM**										
12	Current net rent						£80,000.00				
13	YP for		4 years at		6.00%		3.4651				
14	Term value							**£277,208**			
15											
16	**REVERSION**										
17	Market rental value						£100,000				
18	YP perp.deferred			4 years at		6.00%	13.20156105				
19								**£1,320,156**			
20											
21											
22	**RESULTS:**										
23											
24	**Market Value**							**£1,597,365**			
25											
26	**Less Contract Price**							**£1,597,365**			
27	**Net Present Value**							-£0			
28											
29	**Market Value adjusted for Costs**							**£1,510,510**			
30											
31	**Initial Yield on Market value as a %**							**5.01**			
32											
33	**Yield on Reversion On Market Value as a %**							**6.26**			
34											
35											

Sheet1 / Sheet2 / Sheet3 /

Ready

Leaseholds

The problem of valuing leasehold interests in property has been troubling valuers for many years. In most countries the debate is now closed. However in the UK and in those countries where valuation methodology has largely followed UK practice there is disagreement between those who regard the 'dual rate' (sinking fund) or reinvestment approach to be correct and those who would seek to analyse a leasehold sale as they would any other investment requiring analysis namely to apply single rate principles.

In this chapter the different views are set out with comments. The view of the authors is that the single rate approach is the only objective approach. The responsibility of the valuer is to value according to market behaviour. It will be demonstrated that the only objective analysis that can be made of leasehold sales is to analyse to derive the internal rate of return or equivalent yield. The dual rate approach has been predicated very much on the basis of allowing a set of assumptions as to market behaviour to determine a method of valuation which can then be used subsequently to analyse future sale prices. The valuer may thus be determining the market rather than interpreting the market.

Empirical data collected as part of the RICS research into Property Valuation Methods (1986) indicated that a variety of methods were in use in the market place with a significant 50% using YP Dual Rate unadjusted for Tax. So without positive data to support the authors declared preferences all methods need to be considered leaving the authors to reluctantly advise valuers of leaseholds to identify the most probable purchaser group and as far as possible adopt methods that parallel the investment aims of that most probable purchaser group.

A number of facts are not in dispute, namely:

1. a leasehold interest is of finite duration
2. in investors' terms a lease can only have a market value if it produces an income and if it is assignable. The leasehold income is the profit rent, the rent actually received or potentially receivable less the rent payable to the superior landlord
3. the profit rent must be adjusted for any expenses payable by the interest holder, but irrecoverable from the sub-lessee
4. where tax on income is payable, the capital cost of acquiring a finite investment has to be recovered out of the taxed income

5. the mathematical formulation of the years' purchase single rate may be equated with the years' purchase dual rate

$$\left(\frac{1}{i + ASF} \right)$$

so that capitalisation of an income using either single or dual rate provides the investor with a return on initial outlay plus replacement of capital in a sinking fund, on the assumption of compound interest within the rates used.

The crux of the problem is how the purchaser of a leasehold interest should or does allow for capital recovery. Jon Robinson (1989) in *Property Valuation and Investment Analysis: A Cash Flow Approach* provides a very full summary of the confusions that have emerged over the implicit reinvestment assumptions within the internal rate of return. The book includes the following quotation from the work of Merrett and Sykes:

> Yield, or discounted cash flow return, is correctly defined as the rate of return on the capital outstanding in a project during every year of its life. The interpretation no more assumes that recovered funds are reinvested at the project's rate of return than the statement that a bank is receiving an 8% rate of interest on an overdraft implies that the bank is reinvesting the overdraft repayments at 8% (Merrett and Sykes 1973 p130 in Robinson 1989 p83)

In other words the argument that leasehold investment purchasers need to reinvest in a low safe sinking fund to recover capital and the argument that single rate is inappropriate because leasehold investment purchasers may not be able to reinvest to recover capital at the same rate as the remunerative rate are both non arguments.

In considering the valuation of leasehold investments it is first essential to identify the main categories of leasehold investment.

Categories of leasehold investment

There are a number of identifiable categories of leasehold interests, each of which requires a different approach due to the specific sub-market and profit rent pattern.

1. Medium to long term leaseholds, over (say) 30 years, at a fixed head rent

Where the head lessee is in occupation or has sub-let on a short lease the investment may be treated as one with potential income growth, ie a rising profit rent. Whether a dual rate or single rate approach is followed, the risk rate (remunerative rate) will probably be comparable to that for freehold rack rented properties with similar rent review patterns, but 1–2% higher.

If the head lease is for a term certain in excess of 40 years, then it could be treated as directly comparable to the freehold, because the length of time involved reduces the capital recovery element to a very small percentage of the profit rent, which becomes nearly all spendable income, as in the case of a freehold investment. For example the YP for 40 years at 8% is 11.9246 and the YP in perpetuity at 8% is 12.5 a difference of only 0.58.

Where the head rent is subject to review either in line with any sub lease rent reviews eg at the same time or less frequently, geared directly to the sub lease rents eg simultaneous review to 5% of occupation leases or otherwise e.g. head rent reviewed to a modern ground rent, then the problem of the gearing must be noted and a simple profit rent multiplied by a YP approach may not be an adequate valuation. A split valuation may need to be used as a check e.g. capitalise the head rent and the sub lease rent separately and then deduct the capital values. The more complex problems may well need to be solved using growth explicit DCF methods.

2. Fixed profit rent

Where the head lessee, or predecessor in title, has sub-let the property for the full unexpired term of the head lease without rent review, the profit rent is fixed. The valuation must reflect this fact and the profit rent must be capitalised at an appropriate cost of capital rate. This effectively allows for the depreciating worth of each year's income in terms of purchasing power. A single rate approach is frequently the most effective method. The yield used will be derived from say the redemption yield on gilt edged securities and then adjusted to reflect the added risks of the leasehold in terms of its covenants and liabilities.

3. Occupation leases

If the most probable purchaser is a potential occupier because the lease is being sold with vacant possession or as part of a business transfer then, although an investment valuation may be adopted, it is technically not a leasehold investment. The valuer must take care to see that the valuation reflects the objectives of that type of purchaser. For example, if the lease of retail premises in a prime trading position is to be sold with the benefit of vacant possession then the purchaser will very likely be another retailer. Retailers will not be looking for a property investment return of $x\%$ on the imputed profit rental; they will be seeking a first class outlet for their goods. Their bids are likely to be well above that of most property investors, indeed, analysis of such sales may produce very unusual rates of return whichever way the analysis is undertaken.

Agents in this market may well simply multiply the profit rent by the number of years remaining in which the profit rent is to be enjoyed. The argument is a monetary argument based on the fact that this is the sum which will be saved and so this is the sum that has to be paid to the assignor. Where a lease is being assigned as part of a business transfer then the price paid for the business may include an amount for securing the leasehold profit rent but the price for the business may be based on a multiple of the business profits which are inflated in the short term by the profit rent, hence analysis might be meaningless and endeavours to assess the value of the lease as separate to the value of the business might also be meaningless.

Single rate valuation of leaseholds

All valuations should be based on the analysis of and interpretation of the market. So the first step has to be the analysis of sales of comparable leasehold investments.

Example 7.1

A leasehold investment has just been sold for £294,500. The lease is of retail premises in a popular shopping street. The lease has 20 years to run and the property has been sublet for the balance of the lease without rent review. The profit rent has been assessed at £30,000.

Trial and error analysis is simply the reverse of a valuation. In this case the analysis is to assess the single rate YP used in the valuation which resulted in the sale price OR the analysis is to determine the investment's internal rate of return.

The IRR using a financial calculator is 8.002% for valuation purposes this could be taken as 8%.

Proof

Profit Rent	£30,000
YP 20 years at 8%	9.8181
Market Value	£294,500

There is no way that the sale price of a leasehold investment can be analysed to determine the dual rate objectively unless the analyst makes a prior assumption as to the accumulative rate. This is a first fundamental weakness in the dual rate approach. It can not be based on objective market analysis, it can only be based on the valuers own judgment.

If a single rate approach is to be used in a marketplace where purchasers are liable for tax on incomes then capital recovery must be provided for out of taxed income. Investment advice can be provided on a before tax or after tax basis. In this example the net of tax return with tax at 40% is based on a net income (profit rent) of £18,000 (£30,000 minus tax at 40%). The net of tax return is 1.99%. The reason for this low figure is because of the terminable nature of the investment and because there is no distinction for tax purposes between income which is in effect the return of the initial capital invested and the income which is the genuine investment income (see Chapter 8). To value single rate on a net of tax basis requires the valuer to either, adjust the inherent sinking fund element or deduct tax from the profit rent. The latter approach is adopted in example 7.2.

Example 7.2

Estimate the present worth of a profit rent of £100 for four years. Tax is at 40% and the investor requires a net of tax return of 7.75%.

Profit rent	£100
Less tax at 40p	40
Net of tax profit rent	£60
YP for 4 years @ 7.75%	3.33
	£200

The investor will obtain a return on his/her capital of 7.75% net and will recover his/her capital in full at 7.75%. Whether or not part of the income is reinvested is immaterial, as the investor has the opportunity of accepting a partial return of capital at the end of each year.

	Capital outstanding	Net income at 7.75%	Return on capital	Return of capital
Year				
1	£200	£60	£15.50	£44.5
2	£155.5	£60	£12.05	£47.95
3	£107.55	£60	£8.335	£51.66
4	£55.89	£60	£4.33	£55.67*
				£199.78

*Error due to rounding of figures

Most market valuations are undertaken on a before tax basis so it is essential to have due regard to the net of tax implications of investing in leaseholds.

One of the critical issues that surrounded the debates about single or dual rate was the realisation that many leasehold investments were being purchased by non tax payers while the valuers who were advising sellers of leaseholds were valuing dual rate adjusted for tax. The full impact of this can be reviewed in Baum and Crosby (1995). The point here is that where any market is available to non taxpayers it is the tax payer who suffers and has to accept the poorer net of tax return, hence analysing the market and valuing on a before tax basis would seem to be appropriate for valuation purposes but advising on a net of tax basis is essential for the tax paying investor.

Single rate valuation of leaseholds poses no added issues beyond those already discussed in chapter 5 on freehold valuations. It can comfortably deal with situations akin to the freehold term and reversion without encountering the problems shortly to be noted under the dual rate section.

Example 7.3

Value the leasehold investment in a retail outlet in a popular shopping street. The lease has 20 years to run at a fixed head rent of £50,000. The current sub lease has 5 years to run to the lease renewal at a rent of £80,000 and the Market Rent is £100,000. The appropriate leasehold capitalisation rate is 8% based on market comparables.

Term		
Rent receivable	£80,000	
Rent payable	£50,000	
Profit rent	£30,000	
YP 5 years at 8%	3.9927	£119,781
Reversion		
Rent receivable	£100,000	
Rent payable	£50,000	
Profit rent	£50,000	
YP 15 years at 8%	8.5595	
	£427,975	
PV £ in 5 years at 8%	0.68058	£291,271
Market Value		£411,052

The main issue facing valuers undertaking genuine leasehold investment valuations is the lack of market comparable evidence. Every leasehold interest in property is unique in terms of lease terms so where market analysis is available valuers will frequently find themselves having to make adjustments to overcome differences between the comparable and the subject property. The other valuation issues are the same and require a thorough inspection of the property and documents and clearly the valuation can be based on an in arrears assumption or quarterly in advance. The valuer may even need to deal with situations where the head rent is quarterly in advance and the sub lease rents are monthly in advance. Careful application of basic principles will resolve all these normal issues.

The use of single rate for valuations of leasehold interests avoids the great majority of technical problems about to be identified in relation to dual rate valuations. Net of tax poses no problems using for most purposes 40% as the tax adjustment. The problem of deriving an appropriate leasehold capitalisation rate from a very diverse market remains but is not insurmountable.

It may have been the lack of comparables that persuaded some valuers in the late 19th and early 20th century to switch to dual rate as one of its initial attributes was that it relied on freehold market evidence for deriving the remunerative rate. In considering this and other issues around the subject of dual rate it needs to be remembered that it developed at a time when most leaseholds were long — 99 years — or very long — 999 years ie to all intents they could be treated as freeholds but at a marginally higher capitalisation rate.

Dual rate adjusted for tax

The reasons generally given for valuing leaseholds on the dual rate basis for recoupment (recovery) of capital by reinvestment in a sinking fund at a low safe rate are:

1. that an investor requires a return on initial outlay throughout the lease term at the remunerative rate
2. that all capital must be recovered by the end of the lease term and that to be certain of doing this reinvestment in a sinking fund accumulating at a low, safe, net-of-tax rate must be undertaken
3. that by so doing the investor has equated the finite investment with an infinite (freehold) investment in the same property base, ie, the spendable income is perpetuated.

Example 7.4

Estimate the value of a four-year unexpired lease. The profit rent is £100 pa net.

Net of outgoings profit rent	£100
YP for 4 years at 10% and 3%	
adj tax at 40%	2.00
	£200

Proof:

Annual sinking fund to replace £1 at 3% in 4 years	= 0.239
∴ ASF to replace £200	= 200 × 0.239
	= £47.80

Grossed-up to allow for tax on original income of £100

$$= \quad £47.80 \times \frac{1}{1-t}$$

$$= \quad £47.80 \times \frac{1}{1-0.40}$$

$$= \quad £47.80 \times \frac{1}{0.60}$$

$$= \quad £79.66$$

Grossed-up sinking fund   =   say £80

Therefore :
Spendable income
$$= \quad £100 - £80$$
$$= \quad £20$$
$$= \quad \text{a 10\% return on £200}$$

The £200 recovered over four years can be reinvested at 10% to produce a 10% return in perpetuity.

But note:
1. Capitalisation on a single rate basis can be considered to give a return on initial outlay plus a sinking fund. Additionally it can be shown to give a return on capital at risk.
2. It can be shown that when a single rate years' purchase is used the investor will, at the implied sf rate, recover all capital by the end of the term.
3. The investment may have been perpetuated and can be shown to be perpetuated on a single rate and dual rate basis. But a leasehold interest cannot be equated with a freehold interest if one merely recovers historic cost during periods of inflation.

Within conventional dual rate valuations two schools of thought are apparent. Initially both accept that the risk or remunerative rate for a leasehold should be a minimum of 1–2% above the appropriate freehold rate. This is to allow for the additional problems of reinvesting in a sinking fund, reinvesting the accumulated sinking fund at the end of the lease term, and for the additional risks inherent in a leasehold investment relating to restrictions imposed by the lease terms and possible dilapidations. This philosophy would have been sound in the late 19th and early 20th centuries when leases were for long terms of 99 years or more. To assume some constant magical relationship today is inconsistent with the philosophy of market valuations, unless market transactions justify that rate.

Both schools assume that reinvestment must be at a low, net-of-tax rate and that the sinking fund premium or instalment must be met out of taxed income. But the schools seem to differ on the nature of the sinking fund.

The original theory is that capital replacement in order to be guaranteed must be in effect 'assured'. Hence reinvestment is assumed to be made in a sinking fund policy with an assurance company. Such a policy is a legal contract. The total sum assured will be paid out on the due date if all premiums have been met. The rate of accumulation is assumed to be between 2.5% and 4%. Such a policy is unaffected by any changes in tax rates, as such changes will only affect the investor's investment income. (If a policy is taken out and a re-sale occurs part way through the term, the insured must either continue to pay the annual premium or accept a paid-up policy. In the latter case there would be a shortfall in capital recovery.)

The second school argues in favour of reinvestment at safe rates. These will not necessarily be guaranteed but could conceivably include regular savings in a post office account, bank deposit account, or building society savings or investment accounts.

A potential 'danger' is that the valuer can be implying an effective gross return on the sinking fund greater than the remunerative rate used in a valuation. Conventionally a 6% freehold yield might become a 7% leasehold yield, but 7% and 4% with a tax rate of 50p in the £1 would imply that the investor could invest in a sinking fund and initially be better off than by purchasing the leasehold interest because 4% net is 8% gross. It can be seen that the profit rent must grow at a rate sufficient to compensate for the opportunity cost of investing at 7% gross and not at 4% net.

This would be perfectly tenable under certain market conditions.

Both approaches rely on the argument that the purchaser of a leasehold investment is unlikely to have the opportunity to reinvest at the remunerative rate. Apart from the fact that there is no need to reinvest if the investor accepts a return on capital at risk, there is the obvious point that relatively safe rates, albeit variable, are readily available for regular saving.

Further, purchasers of leasehold investments are not purchasing a single one off property investment but are investing in more than one property and more than one type of investment and may receive sufficient income each year to purchase further leasehold investments. The leasehold investor is then better advised to reinvest in further leaseholds (or freeholds) rather than in a sinking fund.

Historically there may have been some justifications for the dual rate approach. In the 18th, 19th and early 20th century leaseholds had long terms (frequently ground leases or improved ground leases with no rent reviews), and were therefore comparable to freehold investments. Initially they were valued single rate at 0.5–2% above the freehold rate to reflect the additional risk arising from lease covenants etc.

The problem emerged in the 1920s and 1930s when lease length began to fall and when the gap between safe and risky investments became greater. This was in part a reflection of depression in the economy.

It was also a time when purchasers of leases were individuals buying for occupation rather than to hold as investments; there were few large investors in the market. By the mid-1930s valuers were split between single rate and dual rate, although writers still made the point that sale prices could be analysed on either basis. The need to demonstrate parity with freeholds began to dominate, as did the concern for capital recovery when safe investments were yielding so little.

For those who were seeking protection and perpetuation through reinvestment there was the added difficulty of finding reinvestment opportunities for small annual sums other than in a bank or post office. The result was a growing support for using a dual rate approach.

The position post-1960 was very different as inflation became part of life. Nevertheless, the dual rate method survived virtually unchallenged other than for the split between the advocates of non adjusted dual rate and tax adjusted dual rate. It survived in the 1960s, 1970s and 1980s with the low sinking fund even though most savers could achieve much higher returns than the 4% maximum used by valuers. It survived even though few if any sinking fund policies were taken out, even though occupiers were depreciating leaseholds as wasting assets in their accounts and even though investors accepted that they effectively provided for capital replacement by reinvestment in a general sense rather than on a per property basis.

A final point is that where the investment is made with borrowed funds or partially borrowed funds (eg 60% mortgage) reinvestment at the rates adopted in valuations by certain valuers will certainly be at levels lower than the interest charged on the borrowed capital. Clearly the investor would be better advised to pay off part of the loan rather than reinvest in a sinking fund.

One of the most disturbing elements of dual rate valuation practice is its concentration on today's profit rent and its failure to consider the future, in particular its failure to reflect the unique gearing characteristics of longer leaseholds. Taking, for example

- property A held for 20 years at £10,000 pa without review and sub-let with five-year reviews for £20,000 pa, and
- property B similarly held at £80,000 pa and sub-let at £90,000 pa, the conventional dual rate approach would place the same value on both as they both produce a profit rent of £10,000 pa.

This completely hides the fact that with rental growth at any rate per cent the profit rents after each rent review grow at different rates. This problem of gearing can only be solved by valuing the head rent and occupation rent separately and by adopting a DCF or real value approach. Conventional dual rate methodology would, if applied constantly to every situation, provide the investor with some phenomenal investment opportunities and result in some vendors grossly underselling their investments.

The dual rate method was considered in depth in the RICS research report (Trott, 1980) and by Baum and Crosby 1995. A general conclusion at that time was that it would continue to be used because 'that is how the market does it'. The position now is that for many it continues to be used because 'that is what they (the valuers) were taught', because 'it continues to be taught because it has an intellectual challenge', because it seems simple to add 2% to the freehold rate rather than search for comparable leasehold rates.

The authors would maintain that it is defunct because:

1. investors do not seek a return on initial outlay throughout the lease term, they all adopt a portfolio approach and appreciate and use the investment concepts of NPV and IRR
2. investors have never and still do not reinvest in sinking funds
3. investors invest regularly in time limited investments such as short and medium dated stocks and do not seek to equate finite investments with infinite (freehold) investments and do not use dual rate in those markets
4. valuation of leaseholds is based on the analysis of freehold sales or on the basis of analysis of leaseholds based on false assumptions as to tax rates and sinking fund rates
5. valuation of anything other than a fixed profit rent produces added issues for the valuer to overcome if the dual rate approach is to be used.

The following section reviews some of the dual rate issues. The authors offer there apologies to students for the fact that this material is still included in most study materials and UK courses.

Question 1

Value a leasehold profit rent of £20,000 for four years using 8% and 3% adj tax at 40% and prove that a return on capital at 8% and a return of capital at a net rate of 3% is achieved.

Variable profit rents

Variable profit rents present special problems in valuations where the sinking fund rate of return adopted is lower than the investment risk rate (remunerative rate).

The valuation of a leasehold investment may involve a variable profit rent in the following cases:

1. Where a leaseholder lets property to a sub-lessee for part only of the unexpired term. In this case there will be a fixed profit rent for n years and a reversion to a different level of profit rent in n years.
2. Where a leaseholder lets property to a sub-lessee for the full unexpired term with a proviso for rent reviews in n years to a known sum or to full rental value.
3. Where a leaseholder has a right to a new lease under the Landlord and Tenant Act 1954 Part II, and where the rent for the new lease must be at market rental value adjusted for goodwill and/or approved improvements carried out by the lessee (see Landlord and Tenant Act 1954 Part II and Law of Property Act 1969).

Wherever a leasehold valuation is required the valuer should exercise care in determining the true net profit rent, special attention being given to the allowance for expenditure of repairs and insurance when a sub-letting has taken place on terms differing from the head lease.

In some cases the profit rent will be rising, while in others it will be falling. In both cases anomalies occur if, in the assessment of present worth, the flow of income is treated as an immediate annuity plus one or more deferred annuities, assessed on a dual rate basis. The valuation should be treated as an immediate variable annuity.

While the valuer's subjective adjustments to the remunerative rates used in the three cases may well differ, the particular problem of variable profit rents will remain and is fully illustrated in the following problem.

Example 7.5

Using as an example a profit rent of £1,800 for two years rising to £2,000 for five years, demonstrate the problems encountered when valuing rising profit rents on a conventional dual rate basis.

Demonstrate the alternatives to this approach.

Leasehold rate at MRV: 10% and 3% adj tax at 40%

Conventional valuation:*

Term:		
Profit rent	£1,800	
YP for 2 years at 9% and 3% adj tax at 40%	1.0977	£1,976
Reversion:		
Profit rent	£2,000	
YP for 5 years at 10% and 3% adj tax at 40%	2.416	
	£4,832	
PV £1 in 2 years at 10%	0.8265	£3,994
		£5,970

*The use of variable remunerative rates causes further problems as does the deferment of the reversions at the reversionary remunerative rate. These issues can be overcome by adopting a same yield approach.

What are the problems?

None is apparent from this valuation, but compare the valuation of the following profit rent of £1,800 receivable for seven years (five years plus two years):

Profit rent	£1,800	
YP for 7 years at 10% and		
3% adj tax at 40%	3.1495	£5,669.10

Notice that the second investment is valued at £300 lower than the first.

But what is the real difference?

The first investment produces an extra £200 for five years deferred for two years.

Profit rent	£200	
YP for 5 years at 10% and		
3% adj tax at 40%	2.416	
	£483.2	
PV £1 in 2 years at 10%	0.8265	£400.0

The difference in the two valuations is £300, yet the difference in rent is worth £400. There is obviously some kind of error. The error can be demonstrated in the same manner much more dramatically.

Take two profit rents with a 20-year life.

One is of £1,000 for the whole period; the other rises to £1,100 after 10 years (and is therefore more valuable).

	(1)			(2)	
Profit rent	£1,000		Profit rent	£1,000	
YP for 20 years			YP for 10 years		
@ 10% and 3%			at 9% and 3%		
adj tax at 40%	6.1718		adj tax at 40%	4.2484	£4,248
	£6,172		Reversion to	£1,100	
			YP for 10 years		
			at 10% and 3%		
			adj tax at 40%		
			× PV £1 in 10 years		
			at 10%	1.573	£1,730
			(4.0752 × 0.386)		£5,977

The inferior investment is valued more highly by the conventional dual rate method, and the error could be embarrassing because it generally remains unnoticed by most valuation surveyors.

It can be demonstrated, without any need for comparison, that an error does exist by checking that the sinking fund actually replaces the initial outlay. This will be done with reference to the original valuation.

Term
CV   £5,970
Income   £1,800

Spendable income	Sinking fund
$0.09 \times £5,970$	$£1,800 - £537.3$
$= £537.3$	$=$ £1,262.7 gross
	$=$ 0.6 (£1,262.7) net
	$=$ £757.6
A £1 pa for 2 years at 3%	2.03
	£1,538

This £1,538 will then be allowed to accumulate interest for five more years, the period of the reversion.

$$\text{A £1 in 5 years at 3%} \quad \frac{£1,538}{1.1593} \quad = \quad £1,783 \text{ replaced}$$

Reversion:
CV   £5,970
Income   £2,000

Spendable income	Sinking fund
$0.10 \times £5,970$	$£2,000 - £597$
$= £597$	$=$ £1,403 gross
	$=$ 0.6 (£1,403) net
	$=$ £842
A £1 pa for 5 years at 3%	5.3091
	£4,470 replaced

Total replacement of capital £1,783 + £4,470 = £6,253

Compare £6,253 with an initial outlay of £5,970 and it can be seen that there is an over-replacement of capital.

Why?

It can be shown that the replacement of capital for both term and reversion, when examined separately, is perfectly correct.

Term:
CV   £1,976
Income   £1,800

Spendable income		Sinking fund		
0.09 x £1,976		£1,800 – £178	=	£1,622 gross
			=	£178
				× 0.6
				£973.2 net
		A £1 pa for		
		2 years at 3%		2.03
		Term capital replaced		£1,976

Reversion:
CV   £4,832
Income   £2,000

Spendable income		Sinking fund		
0.10 × £4,832		£2,000 – £483.2	=	£1,516.8 gross
= £483.2				× 0.6
				£910.08 net
		A £1 pa for		
		5 years at 3%		5.3091
		Reversion capital replaced	=	£4,832

It follows that the error must arise from the addition of term and reversion. This produces an extra accumulation of sinking fund resulting in an over-replacement of capital. The deferment of the reversion by a single rate PV is another expression of the same error. It is often said that the error results from the provision for two sinking funds or the interruption of the desired single sinking fund. As a result the methods that have been devised to deal with this error attack the problem by attempting to ensure that the initial capital is accurately replaced by the sinking fund.

Method 1: The sinking fund method

The problem which has been identified above is that the conventional dual rate method of valuing leasehold interests does not provide for accurate replacement of capital. The sinking fund method ensures that this must happen, its premise being that capital value is equal to the amount replaced by the sinking fund. The method calculates the amount of the net sinking fund and its accumulation which must necessarily be equal to the capital value of the investment.

Let CV = x

Term:
Rent   =   £1,800
Gross sinking fund   =   Income – spendable income
                                          =   £1,800 – 0.09x

(Return on capital = remunerative rate of 9%: return on capital = 0.09x)

Net sinking fund   =   (£1,800 – 0.09x)(0.6)
                                          =   £1,080 – 0.054x

Reversion:
Rent   =   £2,000
Gross sinking fund   =   £2,000 – 0.10x

Therefore:
Net sinking fund   =   £1,200 – 0.06x

Calculate accumulation of net sinking funds:

Term:		£1,080 – 0.054x	
A £1 pa for 2 years at 3%	2.03		
A £1 in 5 years at 3%	× 1.1593	2.3534	
		£2,541.65 – 0.127x	
Reversion:		£1,200 – 0.06x	
A £1 pa for 5 years at 3%		5.3091	£6,370.29 – 0.31285x
Adding the term and reversion equations together			£8,912.57 – 0.4455x

The capital value should equal the amount replaced and therefore x should equal the sum of the sinking fund accumulations.
   So:
   x =   £8,912.57 – 0.4455x
   1.4455x =   £8,912.57

Therefore
   x =   £6,165.74

This method can be checked by checking the accumulation of sinking funds on term and reversion.

Term:
CV   £6,166
Income   £1,800

Spendable income		Sinking fund		
0.09 × £6,166		£1,800 − £554.94	=	1,245.06 gross
= £554.94				× 0.6
				£747.04 net
A £1 pa for 2 years at 3%	2.03			
A £1 in 5 years at 3%	1.1593			2.354
				£1,758.07

Reversion
CV	£6,166
Income	£2,000

Spendable income		Sinking fund		
0.10 × £6,166		£2,000 − £616.6	=	£1,383.40 gross
= £616.60				× 0.6
				£830.06 net
A £1 pa for 5 years at 3%				5.3091
				£4,406.87

Total Capital replaced (£1,758.07 + £4,406.87) £6,164.94*

*The marginal error here is due to the initial rounding to £6,166 and subsequent rounding in the calculations.

Method 2: The annual equivalent method

The purpose of this second method is to find that fixed income which would be equivalent to the rising profit rent which is to be valued. The current teaching of this and the sinking fund method is attributed to Dr MJ Greaves previously of Reading University and the National University of Singapore.

Equivalent incomes for both term and reversion are found and valued separately to allow for the use of different remunerative rates on term and reversion.

It was originally suggested that the rate of interest used to capitalise and de-capitalise both incomes when finding the annual equivalent should be the accumulative rate, this approach is adopted in this illustration.

A Capitalisation at low safe rate

Term:
Income	£1,800	
YP for 2 years at 3%	1.9135	£3,444

Reversion:
Income	£2,000		
YP for 5 years at 3%	4.5797		
PV £1 in 2 years at 3%	× 0.9426	4.3168	£8,633

B Find annual equivalent income:

Term:

$\qquad$ £3,444 ÷ YP for 7 years at 3%

=$\qquad$ £3,444 ÷ 6.2303

=$\qquad$ £552.78

Reversion:

$\qquad$ £8,633 ÷ YP for 7 years at 3%

=$\qquad$ £8,633 ÷ 6.2303

=$\qquad$ £1,385.65

C Capitalise annual equivalents at market capitalisation rate

Term:

Income	£552.78		
YP for 7 years at 9% and			
3% adj tax at 40%		3.2519	£1,797.59

Reversion:

Income	£1,385.65		
YP for 7 years at 10% and			
3% adj tax at 40%		3.1495	£4,364.10
			£6,161.69

D Proof

Term:

CV	£1,797.59
Income	£1,800

Spendable Income	Sinking fund	
0.09 × £1,797.59	£1,800 − £598.19* =	£1,201.81 gross
= £161.78		× 0.6
		£721.09 net

Reversion:

CV	£4,364.10
Income	£2,000

Spendable income	Sinking fund	
0.10 × £4,364.10	£2,000 − £598.19* =	£1,401.81 gross
= £436.41		× 0.6
		£841.09 net

*Total spendable income at 9% and 10% on term and reversion is £598.19 (£161.78 + £436.41)

Check sinking fund accumulations:

Term:

		£721.09	
A £1 pa for 2 years at 3%	2.03		
A £1 in 5 years at 3%	× 1.1593	2.354	£1,697

Reversion:

	£841.09	
A £1 pa for 5 years at 3%	5.3091	£4,465
		£6,162*

* Error due to rounding.

The proof employed in the sinking fund approach is not applicable in this case, where term and reversion must be kept separate when checking the sinking fund accumulation. In the first approach spendable income differs over term and reversion; but in the annual equivalent method, spendable income remains the same. If the remunerative rates were the same over term and reversion then either proof may be used, but when they differ; the annual equivalent valuation can only be checked by the particular approach outlined above.

Method 3: The double sinking fund method

This, the original of the three methods discussed here, involves more detailed arithmetic. The required sinking fund to replace a capital value of x is deducted from income to leave the amount of spendable income. The spendable income is then capitalised.

This ignores the sinking fund accumulation, which is then added back to produce a similar equation to that which is solved in the sinking fund approach.

A constant sinking fund is ensured by this approach, overcoming the conventional method's fault. This method is attributed to AW Davidson one time Head of Valuation at the University of Reading.

1 Let CV = x Then:
   sinking fund to replace x = ASF to replace £1 in 7 years at 3% × x
   = $0.1305064x$

   Gross up for tax at 40% = $0.1305064x \left(\dfrac{1}{1-t} \right)$

   = $0.1305064 \times x$
   = $0.2175106x$

Therefore:

Spendable income is £1,800 − 0.2175106x for term
Spendable income is £2,000 − 0.2175106x for reversion

2   Capitalise the spendable income

Term:
Income   £1,800 − 0.2175106x
YP for 2 years at 9%   1.7591

                                                                  £3,166.38 − 0.3826228x

Reversion:
Income   £2,000 − 0.2175106x

YP for 5 years at 10%   3.7908
PV £1 in 2 years at 10%   0.8264   3.133

                                                                  £6,265.79 − 0.6814372x
Adding the term and reversion   £9,432.17 − 1.06406x

This (£9,432.17 minus 1.06406x) is the present value of the spendable income provided by the investment. It has been capitalised by a single rate YP which contains an inherent sinking fund at the remunerative rate. The capital value of the spendable income could thus be reinvested at the end of seven years, while another sinking fund has been provided to replace the capital value of the whole income flow — x. There are therefore two sinking funds, hence the name of the method.

An alternative view is to remember that income is split into two parts when it is received for a limited period – spendable income and sinking fund. It follows that capital value can be split into capitalised spendable income and capitalised sinking fund. Having found the first of these constituents the second should be added to give the total value of the investment.

What, then, is the present capital value of the sinking fund?

It replaces x in seven years.

Its present value is x deferred for seven years at the investor's remunerative rate(s): 9% for two years and 10% for the remaining five.

3   Replaced CV = x

PV £1 in 5 years at 10%   0.6209213
PV £1 in 2 years at 9%   0.84168   0.522617

                                                                  0.522617x

Thus if x is the capital value CV, then x must be equal to the total of 1, 2 and 3.

x	=	£9,432.17 − 1.06406x + 0.522617x
x	=	9,432.17 − 0.541443x
1.541443x	=	£9,432.17
x	=	£6,119.05

Proof:
Term:
CV   £6,119.05
Income   £1,800

Spendable income   Sinking fund
0.09 × £6,119.05   £1,800 − £550.71   =   £1,249.29 gross
= £550.71   × 0.6
                                                  £749.57 net

Reversion:
CV   £6,119.05
Income   £2,000

Spendable income   Sinking fund
0.10 × £6,119.05   £2,000 − £611.91   =   £1,388.09 gross
= £611.91   × 0.6
                                                  £832.85 net

Check sinking fund accumulations

Term: £749.57
A £1 pa for 2 years at 3%   2.03
A £1 in 5 years at 3%   × 1.1593   2.354   £1,764.02

Reversion: £832.85
A £1 pa for 5 years @ 3%   5.3091   £4,421.68
                                                   £6,185.70
(compare with CV of £6,119.05)

The proof suggests that capital value is not accurately replaced. But the rationale of the method must ensure accurate replacement of capital, and this leads to the conclusion that this method suffers from another fault. This is that although rates of return of 9% on the term and 10% on the reversion are required they are not accurately provided by this valuation.

There is an over-replacement of capital and consequently the interest is undervalued, as in the conventional method, but to a reduced degree.

	CV	*Replaced CV*
Conventional dual rate	£5,970	£6,253
Sinking Fund	£6,166	£6,165
Annual equivalent	£6,162	£6,162
Double sinking fund	£6,119	£6,186

It is therefore possible to conclude that the sinking fund and annual equivalent methods best overcome the problem posed by the conventional method of valuing variable profit rents as they appear to provide for accurate replacement of capital and correctly allow for the required rate of return to be provided.

However, even the apparent accuracy of these solutions cannot be relied upon in all circumstances.

For example, the sinking fund method becomes unworkable if the term income is particularly low; the spendable income over the period becomes negative, and considerable problems arise.

Harker, Nanthakumaran and Rogers (1988) of Aberdeen University in their discussion paper on 'Double sinking fund correction methods' support the sinking fund method and suggest a resolution to the negative spendable income problem.

A popular alternative to these methods is the Pannell method. In this method the capital value of the variable profit rents are found by capitalising each on a single rate basis using the appropriate leasehold remunerative rates. The annual equivalent of the product is then found by dividing through by the YP for the full unexpired term at the remunerative rate or the average of the rates used. This annual equivalent can then be capitalised on a dual rate basis for the full term.

Because the use of dual rate valuations can be seen to lead to difficulties, single rate leasehold valuations are recommended. The fact that few valuations of variable profit rents are actually valued using one of the corrective approaches raises other questions about the validity of using dual rate. It should be noted that if single rate is used the valuations may be prone to error where more than one rate of interest is used (see p103).

None the less, the potential for arithmetical error in a single rate valuation is considerably less. A reasonable conclusion to this discussion on dual rate is that if valuers are supposed to mirror the market in the selection of valuation methods then it is reasonable to suggest that if purchasers in the market do not use dual rate then neither should valuers. Most of this chapter is therefore redundant.

Question 2

A shopping centre currently produces rents of £2.5 million with rents due for review in three years to £3.25 million and thereafter every five years. The centre is held leasehold with 60 years of the lease to run with one rent review of the ground rent in ten years time. The ground rent is currently £50,000 a year and the rent review is fixed at 5% of the shop rentals.

Set out skeleton valuations of the head lease using a dual rate approach at 8% with tax at 40p and a DCF approach on the basis of a freehold target rate of 10%.

Tax adjustment rate

What rate of tax should be used to adjust a gross profit rent to a net rent, or to gross-up a sinking fund to allow for tax?

The answer normally given is to use standard rates of income or corporation tax but the problem is more complex due to the presence of non tax payers such as pension funds and charities. The valuer's task is generally to determine market value, this implies a sale to the most probable purchaser, which implies some knowledge of the market and therefore sufficient knowledge to adjust, both for analysis and valuation, at a rate appropriate to the most probable purchaser.

However, as tax is levied at different rates for different investors, an average tax rate (eg 40%) is used to reflect market interaction. But analysis of leasehold investments from a client's point of view must be carried out at that client's net-of-tax rate.

Example 7.6

'Short leasehold investments are sound investments for charities because of their tax advantage.' Discuss.

Apart from the point that charities are more interested in income over the long term than short term, a number of points can be made. If the statement is true then charities must comprise a sub-market for this kind of investment. In that case the valuer needs to reflect the fact that the income is probably tax free; that they would be ill-advised to reinvest in a sinking fund policy with an assurance company because they will probably have difficulty in recovering the tax element on the 4% net accumulations of the policy; and that they would be best advised to reinvest in the safest gross funds* to avoid delay in recovery of taxed interest or dividends.

* The term 'safe gross funds' is used here to indicate any investment where interest or dividends are paid without deduction of tax.

Example 7.7

Analyse the sale of a £100 profit rent for 4 years at a price of £200 from the point of view of a gross (ie non-tax-paying) fund.

£200/£100 = multiplier of 2

(a) Assume reinvestment in an equivalent safe gross fund at, say, 8%. Therefore the sinking fund to recover £200 will be:

£200 × ASF to replace £1 in 4 years at 8%
= £200 × 0.22
= £44

The gross spendable income   = £100 − £44
                                     = £56

Therefore

The gross return   = $\dfrac{56}{200}$ × 100 = 28%

(b) Without a reinvestment assumption analyse to find the internal rate of return.

Solve $\dfrac{1 - \left(\dfrac{1}{(1 + i)^4}\right)}{i}$ = 2

The present value of £1 pa in 4 years at 35% = 2, and therefore the IRR is 35%.

Charities, because of their tax position, may be interested in short leasehold investments, because, if they can buy at prices determined by conventional dual rate approaches, the effective return will be sufficiently high to compensate for the disadvantages of the investment.

Summary

The historical evolution of the dual rate method is now shrouded in the mists of time. Textbooks and journals of the time move from single rate to dual rate with very little explanation and with virtually no consideration of the fact that it hinges on the acceptance of a return on initial outlay throughout the life of the investment, a concept apparently unique to the UK leasehold property investment market.

The emergence of the tax adjustment factor is better documented and is relevant to all short-term investments, however valued, where tax is charged on that element of income which is essentially capital recoupment.

Part of the market is still as reluctant today to return to single rate as it was initially to shift from single rate to dual rate. A strength of dual rate lay in the ability to compare returns from freeholds with leaseholds. The reluctance today may be the fear of the unknown arising from the unique nature of every leasehold investment and the pressure on the valuer to find a unique solution. The solution lies in the proper use of discounting techniques where the unique growth expectations can be explicitly accounted for.

In summary dual rate methodology can be criticised because:

- sinking funds per se are not available in the market
- leaseholds as investments are not directly comparable to freeholds and remunerative rates cannot be taken as 1–2% above freehold rates
- accumulative rates do not appear to be market sensitive. 4% has been used when safe rates have been as high as 12% and when they have been below 4%
- it is not possible to derive a dual rate adjusted for tax from market sales unless at least two of the three variables are assumed by the valuer (tax rate, SF rate and/or remunerative rate)
- dual rate analysis of variable geared profit rents is similarly impossible
- there is an arithmetic error in dual rate when used for valuing variable profit rents.

Taxation and Valuation

8

It is customary in valuation to ignore the effects of income tax, on the grounds that investors compare investments on the basis of their gross rates of return. This may well be an acceptable criterion where tax affects all investments and all investors in a like manner. Although this is not the case in the property market only a few valuers would argue that tax should always be deducted from income before being capitalised at a net-of-tax capitalisation rate a 'true net approach'.

One of the most important points to note is that where the income from property is all return on capital (that is, true spendable income), gross and net valuations will produce the same value estimate. Where part of the income is a part return of capital this, other than in the case of certain life annuities, is not exempt from taxation and in such cases the gross and net approaches may produce a different value estimate.

In addition, certain investments will produce fairly substantial growth in capital value over a relatively short term, due to a growth in income. In these cases, if the investment is resold, Capital Gains Tax (CGT) may be payable on the gain realised.

Example 8.1

Explain what is meant by 'net rates of interest' and discuss their uses in valuation, using numerical examples.

A net rate of interest is interest earned on deposited monies after the deduction of an allowance for the payment of tax on the gross interest earned. Here the phrase refers to any net-of-tax rates.

Thus £1,000 deposited with a bank, for example, earning interest at the rate of 10% pa would, with tax at 40p in the £(40%), produce a net rate of 6% ($10 \times (1 - t)$) where t is 0.4. As far as the valuer's use of the valuation tables is concerned, the switch from gross to net valuations should cause no problems. Thus, the present value of £1 in 5 years at 10% gross allowing for tax at 40p in the £ can be found in tax-adjusted tables to be 0.74726, or in unadjusted tables at 6% ($10 \times (1 - t)$) where t is 0.4 to be 0.747258. It should be obvious that where £100 is invested at 10%, and tax is payable, compounding can only take place at the net-of-tax rate.

It is also obvious that, where an investor is a taxpayer paying tax on all investment income at 40%, the actual return after tax will be reduced. The significance of this is not lost on investors, who always have full regard to their net-of-tax returns. When considering property investments, it is therefore necessary to consider whether valuations should also be based on net-of-tax incomes and yields.

For simplicity of illustration a tax rate of 50% is used in some of the examples.

Incomes in perpetuity

The formula for capitalising an income in perpetuity is income divided by *i*.
   If the gross income is £1,000 and the gross capitalisation rate is 10% then:

$$£1,000 \div 0.10 = £10,000$$

If tax is payable at 50p in the £ then

$$£1000 \times \left(\frac{1-t}{1}\right) \div 0.10 \times \left(\frac{1-t}{1}\right) = 500 \div 0.05 = £10,000$$

Clearly as the numerator and denominator are multiplied by a constant $\left(\frac{1-t}{1}\right)$ then they can be divided through by that constant.

Hence:

$$I \div i = \left(\frac{I \times \left(\frac{1-t}{1}\right)}{i \times \left(\frac{1-t}{1}\right)}\right)$$

No difference in value estimate will occur, because the income is perpetual and is all return on capital.

Finite or terminable incomes

An income receivable for a fixed term of years may have a present value which can be assessed on a single or dual rate basis, gross or net of tax.
   Consider an income of £1000 receivable for five years on a 10% gross basis with tax at 50%.

A Single rate

	Gross		Net of tax at 50p	
	£1000			£500
PV £1 pa for 5 years at 10%	3.7908		PV £1 pa for 5 years at 5%	4.32955
	£3,790			£2,164

Here there is a clear difference between the two figures which can be seen to result from the tax adjustments that have been made.

$$I \times \left(\frac{1 - \frac{1}{(1 + i)^n}}{i} \right)$$

cannot be equated with

$$I(1 - t) \times \left[\frac{1 - \frac{1}{(1 + i(1 - t)^n}}{(1 - t)} \right]$$

Here, part of the £1,000 income is a return of capital. If tax is payable at 50p in the £, then only £500 is available to provide a return on capital and a return of capital. A purchase at £3,790 would be too high to allow for the returns implied if tax is at 50p. Where tax is payable, replacement of capital (purchase price) must be made out of taxed income.

In order to preserve the gross rate of 10% for investment comparison the valuation can be reworked, recognising that the return on capital is at 10% and the return of capital is at a net rate out of taxed income. i is therefore 0.10 and the ASF in the formula is at 5% adjusted for tax.

$$£1000 \times \frac{1}{0.10 + \left(0.1809 \times \frac{1}{1 - t}\right)} = £2165$$

or, set out as a valuation:

	£1,000
YP for five years at 10% and 5% adj tax at 50%*	2.165
	£2,165

$$\text{Formula} = \frac{1}{i + ASF} = \frac{1}{0.10 + \left(0.1809 \left(\frac{1}{1 - t}\right)\right)} = \frac{1}{0.10 + 0.3618} = 2.165$$

The gross valuation has been adjusted to equate with the net valuation. Where there is a tax liability, the net valuation is deemed to be more correct because capital cost must be recovered out of taxed income.

Capital outstanding		10% gross	Capital recovered
			(£1,000 − 10% − 50p in £) tax
1	2,165.00	216.500	391.75
2	1,773.25	177.325	411.3375
3	1,361.91	136.190	431.90
4	930.01	93.000	453.50
5	476.50	47.650	476.175*
			£2,165

* Error due to rounding

B Dual rate

The realisation that capital cost may have to be recovered out of taxed income has, for many years, been recognised in the valuation of leasehold property and has resulted in the production of dual rate tax-adjusted tables. Assuming that £1,000 is the profit rent produced by a leasehold interest, and adopting the same gross yield of 10%, a sinking fund at 3% and tax at 50%, the following gross and net valuations may be made.

	Gross
	£1,000
YP for 5 years at 10% and 3% adj tax at 50%	2.098
	£2,098

	Net
	£500
YP for 5 years at 5% and 3%	4.196
	£2,098

The net valuation is of an income of £1,000 less tax at 50%, namely £500. With a gross rate of return of 10%, the net rate must be 5% and, as the sinking fund rate is also assumed to be net, allowance has already been made for tax on the sinking fund accumulations. As the whole income has been netted no further tax adjustment is needed.

The gross valuation becomes in effect a net valuation, because it assumes both a net accumulative rate and a grossing-up factor to allow for the incidence of tax on that part of the income representing capital recovery.

Thus in two cases — incomes receivable in perpetuity, and leasehold interests where dual rate net of tax is used — the normal valuation approach has been shown to be equivalent to the true net approach, apparently obviating any dilemma. However, problems arise when considering deferred incomes, or cases where property is let below market rental value.

Deferred incomes

A deferred income is one due to commence at a given date in the future.

Example 8.2

Calculate the present worth of an income of £2,000 pa in perpetuity due to commence in five years' time. A 10% gross return is required, and tax is payable at 50%.

	Gross			Net
	£2,000			£1,000
YP perp at 10%	10	YP perp at 5%		20
	£20,000			£20,000
PV £1 in 5 years		PV £1 in 5 years		
at 10%	0.6209	at 5%		0.7835
	£12,418			£15,670

Again, it can be seen that capitalisation in perpetuity gross and net each produce the same value of £20,000 in five years time, but, by further discounting to allow for the five years' deferment, the gross and net valuations of the deferred income produce different results arising from the deferment factors of 0.6209 and 0.7835.

Obviously a purchaser would wish to pay only £12,418, *but what of the vendor?*

Assuming initially that capital gains are not taxable, one can see that an investment purchased for £12,418 held for five years and sold for £20,000 is a 10% investment.

£12,418 × amount £1 for 5 years at 10% = £20,000

But if an investor were to deposit £12,418 in an income-producing investment, the 10% return of £1,241.80 in year one would be taxed. From a taxpayer's point of view, a 10% return, all in capital growth with no tax would be better than 10% all in income subject to taxation.

If there was no Capital Gains Tax (CGT) capital growth investments would be very attractive to high rate taxpayers, to such an extent that prices could be pushed up to the point of indifference, ie until the capital growth investment is equated with an income investment. In this example, if tax is at 50p in the £, a taxpayer would get the same return after tax from depositing £15,670 at 10% gross as from purchasing £2000 pa in perpetuity deferred five years for £15,670.

The introduction of CGT in 1965 reduced the tax-free element of capital growth but did not lessen the belief held by some valuers that the net approach was still correct. It is recognised that the incidence of all taxes should be reflected in an investor's true return. Therefore, if tax is material to investors' decisions in the marketplace it should be allowed for in a valuation.

To explain a need for net valuation, a single question needs to be posed.

If investors are assured of a 10% gross return and pay tax at 50%, are they expecting a 5% net-of-tax return?

To this one could add a supplementary question.

Are investors interested in their return net of tax?

It must be assumed that as the Inland Revenue has a prior claim then investors must have some regard to the net-of-tax income. Such an assumption does not infer that valuations must be undertaken on a net of tax basis. Current market practice appears to be to value gross but, if requested, to advise clients, in consultation with their tax advisors, on a net of tax basis.

Rising freehold incomes

Combining the £1,000 income for five years with the £2,000 income commencing in five years and continuing in perpetuity, the result will be a normal term and reversion valuation.

Using the figures from the preceding examples, the gross and net valuations produce these results:

	Gross	Net
Value of £1,000 for 5 years	£3,790	£2,164
Value today of reversion to £20,000	£12,418	£15,670
	£16,208	£17,834

(In such cases, the net valuation will always be higher than its gross equivalent.)

The gross valuation of £16,208 may be analysed in two parts: an expenditure of £3,790 to acquire £1,000 for five years coupled with an expenditure of £12,418 for £20,000 in five years' time. In the absence of income tax, the investor readily achieves his 10% return: but what of his 5% net return?

After tax on income, the investment will have the following cash flow:

Year	Cash flow
1	£500
2	£500
3	£500
4	£500
5	£20,500

This is the cash flow assumed for the 5% net valuation, and the investor would achieve a 5% return at an acquisition cost of £17,834. Therefore at £16,208 the investment must be showing a better return than 5%. If investors only expect a net return of 5%, valuing gross may result in vendors' interests being undersold.

Gross funds

The term 'gross funds' is used today to describe any investor exempt from income, corporation and/or capital gains tax. These are principally charities, local authorities, approved superannuation funds, friendly societies and registered trade unions. Other institutions, such as insurance companies, enjoy partial relief by being assessed at a lower rate. Many first-class property investments are now held by gross funds, which are an increasingly large part of the property market — so much so that co-ownership schemes in agricultural investment and in commercial investment have been carefully created to allow small and large funds to buy, as co-owners, substantial single investments in property while still preserving their tax-exempt status.

If the most probable purchaser in the market is a gross fund then the sub-market is one of gross funds. Because they pay neither income nor capital gains tax, the return to them will be both their gross and net return. Valuations within this sub-market must be based on analysis of comparable transactions, which suggests that no adjustments should be made for tax.

If the current government proposals to introduce Real Estate Investment Trusts (REITs) materialises with the proclaimed opportunity to invest on a gross basis with tax being paid by investors only on their dividends, then a further substantial group of investors will enter the market as gross funds which would further support the continuation of a gross approach to property valuation.

Net or gross?

Because a difference in the valuation may result when valuing net or gross of tax, the problem therefore remains.

Should the valuation be net or gross?

There is no categorical answer: a number of points may, however, be made.

1. In certain cases a net valuation will not produce an accurate estimate of market value, as every potential purchaser may have a different tax liability.
2. A net valuation should be carried out when advising a purchaser of his maximum bid for an investment when his required net-of-tax return is known.
3. In certain sub-markets the market is dominated by non-taxpayers. In these cases a gross valuation will produce the same valuation as a net valuation allowing for tax at 0%.

Capital Gains Tax (CGT)

The introduction of CGT in 1965 had an immediate impact on the property market. Initially it reduced the obvious benefit of investing in properties with high capital growth expectations such as freeholds with early substantial reversions. The market soon adapted to the changes but although there were pleas from a few to move to a net-of-tax approach to valuation the market responded with a simple adjustment to the ARY to account for this new risk element.

Reversionary properties still remained more attractive to individuals because of the then differences between the CGT tax rate and the very high personal tax rates.

The introduction of indexation provisions in the Finance Act 1984 and the revision of those provisions in the Finance Act 1988, whereby only gains in excess of the indexed (retail price index) gains were to be taxed, increased the attractions of capital growth investments. Subsequent changes to tax rates and the provision to tax gains at the same rate as that for taxing a taxpayer's highest income have eroded the advantages of growth investments over income investments.

The remaining tax advantages are:

1. CGT is charged on net gains, ie after deduction of purchase and sale costs and allowable expenses.
2. CGT is charged on the gain after adjustment for indexation.
3. CGT is charged on the gain after deducting the taxpayer's tax free allowance.
4. The tax demand may be up to two years after the gain is realised.
5. CGT may, in some cases, be deferred through roll over relief.

For all these reasons individuals with a 40% tax liability may still favour the highly reversionary properties in preference to rack rented properties. But it is probably the ability of the right property to outpace inflation that attracts the private investor.

For the major investors the changes in indexation provisions have made it easier to take decisions to sell than previously. But because property is normally purchased as a long-term investment any gain is likely to be deferred many years and will not therefore have a significant bearing on purchase price. In addition, 'rollover-relief' may be possible, effectively postponing any CGT and, in any event, for certain classes of investor their exemption from CGT makes them a market in their own right.

There may, however, be obvious cases where property is to be purchased with the intention of realising a capital gain at a specific date in the future. Where the purchasers in the market are likely to have the same intent and the client requires advice on gross and net of tax returns it might be necessary to reflect CGT in the valuation. The client's tax consultant should be involved in any net of tax advice offered by a valuer.

Value Added Tax (VAT)

Many aspects of developing, owning, running and maintaining property including property transactions, other than exempt categories of property, incur VAT. This functions as a significant tax in those cases where it has to be paid and the payer is in a non recoverable position. The assessment of market value is generally undertaken without specific regard to VAT. Where VAT may need to be considered is in the area of property investment advice where buying or letting on the wrong basis regarding VAT could be very costly. Details of this subject are outside the scope of this book.

Summary

- Income capitalisation of property is usually undertaken on a before tax basis.
- Tax on income has no effect where all income is return on capital.
- Tax may need to be reflected where investments are terminable in the short term and tax is a factor in the market place.
- Tax is potentially significant when valuing leasehold investments in a market dominated by tax payers.

Landlord and Tenant Valuations

The income approach to property valuation centres particularly on the relationship between landlord and tenant. This chapter examines several of the valuation problems raised by this relationship.

Premiums

A premium is a lump sum paid by a tenant to a landlord in consideration for a lease granted at a low rent, or for some other benefit. 'At a low rent' signifies a rent below market rental value, and the other benefits will be as a rule, financial, having the same effect as a reduction in rent.

Examples of this are the tenant paying for repairs that would normally be the landlord's responsibility, and the tenant financing the extension of the property without being charged an increase in rent.

A premium is often paid on the grant or renewal of a lease, but there may be more than one premium, payable at any time during the lease term. It entails a cash gain coupled with a loss of rent for the landlord because the usual result of charging a premium will be a letting at less than market rental value. The landlord is therefore selling part of his income.

The tenant will be paying a lump sum in return for a lease at a rent below market rental value, effectively buying a profit rent.

This can be illustrated by looking at the relationship between market rent and market value. Given a market rent of £10,000 and a market capitalisation rate of 10% the market value would be £100,000. A sale represents the full disposal by the owner of all rights to any part of the market rent. But if the owner only wished to sell part of their entitlement they could effectively sell five, ten, 15 or however many years of their freehold title to all of the property or they could sell part of their right to the £10,000 a year rent. A premium is in effect a part disposal to a tenant.

The following table illustrates the premium an owner would require in lieu of an increasing reduction of market rent over a five year term.

Reduction in rent	YP for 5 years at 10%	Premium required at 10% to nearest £1
£1,000	3.7908	£3,791
£2,000	3.7908	£7,582
£3,000	3.7908	£11,372
£4,000	3.7908	£15,163
£5,000	3.7908	£18,954

The explanation is simple in that for every £1,000 in rent reduction over the five year term the market value of the owner's freehold interest falls by £3,790.80 rounded to say £3,791.

Why pay a premium?

The payment of a premium has many advantages to a landlord, and few to the tenant. It could be concluded that premiums will usually be paid where there is a seller's market, ie where there is competition among prospective tenants to secure an agreement with the prospective landlord.

The advantages to a landlord are several.

1. Although the amount of the premium will reflect the discounting of future income, an immediate lump sum receivable instead of a future flow of income is often more attractive due to the 'time value of money'. The landlord may prefer a lump sum in order to meet an immediate expense or to make any kind of cash investment.
2. Receipt of a lump sum immediately may reduce the diminishing effect that inflation has on the value of future income in real terms, especially if rent review periods are longer than is favourable to the landlord.
3. A premium may have tax advantages, but these are substantially reduced now that income and capital gains are taxed at the same rate.
4. A premium should increase the landlord's security of income. Once the tenant has paid a premium, he has invested money in his occupation of the premises in expectation of making an actual or notional profit rent. As a result he is more likely to remain in occupation of the premises and should be a more reliable tenant. Some of the risk attached to the investment from the landlord's point of view may be reduced.

The advantages to a tenant are less well-defined.

1. A premium may be useful as a loss or deduction to be made from profits when being assessed for income tax or capital gains tax.
2. Paying a premium may be advantageous to a tenant when his/her financial circumstances are suchthat he/she prefers to part with capital in order to reduce future recurring expenses.

However, the landlord will usually enjoy the greater benefits. When a property attracts many prospective tenants a landlord may demand and receive a premium in addition to rent. Valuers must be careful to note that this represents the capital value of the extra rent, above market rent, that such excess competition can sometimes generate. The following section deals with the assessment of premiums for rents forgone.

Valuation technique

A premium entails a loss to the landlord of part of the income and a gain to the tenant of a profit rent. The amount of the premium should be calculated so that each party is in virtually the same position as if full rental value were to be paid and received.

The gain/loss of rent, capitalised over the period for which it is applicable, should be calculated to be equal to the amount of the premium. It is conventional practice to use full freehold rates to capitalise the landlord's loss of rent and full leasehold rates to capitalise the tenant's gain of profit rent, presumably on the basis that each party by definition is to be in the same position as if market rental value were being paid and received. Such a conventional approach from a leaseholder's point of view can be criticised where a valuer seeks to use a dual rate or dual rate adjusted for tax approach. The whole argument put forward for using dual rate is based on a false premise about capital recovery. It needs to be understood that where a premium is being paid in lieu of rent it is simply a time value money exchange and the calculations should only be undertaken on a single rate basis. The following examples illustrate typical negotiation stances that might be encountered in the market where the tenant's advisors are basing their advices on dual rate methodology.

The following three examples assume :

- a freehold rate of 10%
- leasehold rates of 11% and 3% with tax at 40p
- a market rental value of £2,500
- premiums payable at the start of the lease and no rent reviews.

Example 9.1

What premium should A charge on the grant of a 21 year lease to B at a rent of £1,500? The market rent is £2,500.

Therefore :

A's loss of rent =	£1,000	
YP for 21 years at 10%	8.6487	£8,649

Example 9.2

A agrees to grant the lease in 9.1. What premium should B offer?

Market rental value	£2,500
Rent agreed	£1,500

Therefore:

B's gain of profit rent =	£1,000	
YP for 21 years at 11% and		
3% adj tax at 40%	5.9481	£5,948

The problem of valuing freehold interests single rate and leasehold interests dual rate is here presented. In practice negotiators tend to suggest that a compromise sum should be agreed by adopting the following approach:

1   Average the final offers:
     (£8,649 + £5,948)/2 = £7,298.5 say £7,300

2   Average the YPs:
     (8.6487 + 5.9481)/2 = 7.298 × difference in full and agreed rents of £1,000= £7,298

In most cases the tenant is due to take occupation and the notional nature of the profit rent is not taxable, and so some valuers will adopt a gross approach to the leasehold calculation bringing the figures much closer together. Thus £1,000 × YP 21 years at 11% and 3% = £1,000 × 6.9027 = £6,903.

However the moment the lease is signed the market value of the freehold interest will fall by £8,649 and so there is no justification for any valuer advising a freeholder to recommend, accepting in lieu of rent, anything less than the amount which equates the market value of their interest, where the tenant is paying the market rent, with the market value of their interest at the lower rent plus the premium. The only ground for doing so would be non financial.

Example 9.3

A is to grant B a lease for 28 years. A premium of £18,000 is to be paid.

What rent should be agreed between the parties?

Again advisors appear to be willing to negotiate on a compromise basis but this will potentially put the freeholder into a poorer financial ie a poorer market value position.

Taking the average of two YP figures can save an unnecessary stage in the calculations but actually hides the potential loss to the freeholder. The approach is as set out below but a before and after valuation of the freeholder's interest in the property will support the argument that the freeholder should only consider the position on a single rate basis and a freeholder's advisors should not be lulled into accepting a deal based on a dual rate or average of two rates approach.

	A	B
Market rental value	£2,500	£2,500
Agreed rent	x	x
YP 28 years	9.3066 (10%)	6.7194 (11% and 3% adj tax at 40%)

Average YP (9.3066 + 6.7194)/2 = 8.013

Then 8.013 (£2,500 − x)	=	£18,000
£20,032.5 − 8.013x	=	£18,000
8.013x	=	£2,032.5 (£20,032.5 − £18,000)
x	=	£253.65 pa

The gain or loss of rent is (£2,500 − x). When capitalised by using an average YP, this is the average capitalised gain or loss of rent and is thus the value of the premium. The unknown, x, can then be calculated. If the freeholder agrees to this then they will be accepting a financial loss.

As has already been suggested, a tenant's repair, improvement or extension may be treated as a premium.

Example 9.4

A agrees to grant to B a 21-year lease at a rent of £12,000 pa which is below market rental value. B has to repair the property at the commencement of the lease as part of the contract at a cost of £15,000.

Estimate the full rental value.

	A	B
Market rental value	x	x
Agreed rent	£12,000	£12,000
YP 21 years	8.6487 (10%)	5.9481 (11% + 3% tax at 40p)
Average YP	(8.6487 + 5.9481)/2 = 7.2984	

Then:

7.2984 $(x - £12,000)$	=	£15,000
7.2984x − £87,581	=	£15,000
7.2984x	=	£102,581
x	=	£14,055 pa

(The cost of repairs is in the nature of a premium for which B expects a reduced rent. $(x - £12,000)$ is the gain/loss of rent, and multiplying this by the average YP gives the average capitalised gain or loss of rent.)

This figure of £14,055 must be the market rental value of the premises on these assumptions. This was never a negotiated figure, as it was a known market fact borne in mind by both parties during the negotiations. But the use of the average YP is necessary as the rent of £12,000 would have been arrived at by negotiation.

While this is a useful theoretical analysis the reader is reminded that it is a very poor way of establishing market rental value and is generally less acceptable to the courts or an arbitrator.

The conventional investment method can give rise to answers that are hard to accept or indeed should be unacceptable to freeholders. In examples 9.1 and 9.2 A required £8,649 in compensation for the loss of rent. B could only afford to offer £5,948, as this was all the gain of profit rent was worth. An agreement at £7,300 would thus make both parties unhappy. But it must be remembered that dual rate valuations using a remunerative rate 1% higher than for freehold valuations inevitably cause lower valuations of the interest, and that dual rate valuations are net in a gross form.

Another approach is the 'before and after valuation'. From the freeholder's point of view, whenever a lease at a low rent is granted the market value of the property may be depreciated by a sum greater than the initial premium calculation suggests. In example 9.1 £8,649 is the premium but the payment of £8,649 alters the nature of the investment from a growth income of £2,500 to a fixed income of £1,500 for 21 years with a reversion to FRV in 21 years' time.

Thus the respective capitalisation rates may differ. If this is the case, a 'before and after valuation' is needed.

Before		
MRV		£2,500
YP perp at 10%		10
		£25,000

After

Rent reserved	£1,500	
YP 21 years at 12%*	7.562	£11,343
Reversion to	£2,500	
YP rev perp after 21 years at 10%	1.35	£3,375
Estimated capital value		£14,718

£25,000 – £14,718 = Premium of £10,282

* Cost of capital or money rate as income is fixed for 21 years

A further problem to be overcome is that the freeholder has indeed lost precisely £1,000 a year but the tenant has gained a rising profit rent starting at £1,000 a year. Traditional valuation methods do not fully reflect this factor. It seems reasonable to question the use of dual rate valuations in such problems and in the valuation of leasehold interests in general (see Chapter 7). A truer picture of real gain and real loss requires an explicit DCF approach.

Emerging in the 1990s was the concept of reverse premiums. Typically an owner of property reluctant to appear to be letting the whole or part of a property below historic market rental levels would offer a premium to a tenant on taking a lease. Similarly tenants occupying over-rented property have had to offer premiums when sub-letting the whole or part. The key to the valuation problem is to recognise that a reverse premium is a payment from a landlord to a tenant. The principles behind their calculation are similar and a before and after valuation is recommended to reflect the market attitude to properties when the rent is being held artificially above normal market rents.

Question

Calculate the premium to be paid when shop property (5% ARY) is to be let at £50,000 on 5-year normal lease terms when the market rental value on similar terms is £75,000.

Future costs and receipts

A study of conventional techniques of deferring future costs raises a number of further questions which are examined shortly.

The deferment of future receipts is relatively straightforward. Often an investment will provide a capital sum at some given time in the future. An example of this is a premium payable during the currency of a lease. This is part of the investment and could therefore be discounted at the remunerative rate.

Example 9.5

A lets 10 High Street to B for 21 years at £13,925 pa. In addition, a premium of £15,000 is payable at the end of the lease. FRV is £15,000 pa. Value A's interest.

Assuming a freehold rate of 10%

Rent	£13,925	
YP for 21 years at 9%	9.2922	£129,393
Rent	£15,000	
YP rev perp after 21 years at 10%	1.35131	£20,269
Premium	£15,000	
PV £1 in 21 years at 10%	0.13513	£2,027
Estimated capital value		£151,689

This approach implies that future fixed receipts should be discounted at the remunerative rate and added to the freehold value. In which case the premium of £15,000 varies in value according to the quality of the investment. This conventional approach cannot be accepted. The present worth of any known future sum must be found by reference to money market rates.

Future capital costs will often arise out of investment in property. Conventional valuation practice distinguishes two types of future capital cost.

1 Liabilities

These may have to be incurred for some reason and are a cost to the investor, often being legally enforceable. They must be allowed for in a valuation. Examples are premiums to be paid in the future; a sum to be spent on dilapidations at the end of the lease; or work that must be carried out under fire regulations, etc.

The investor must make certain that the cash required for the liability is available at the relevant time. Any accumulation must therefore be risk-free, and it follows that conventional practice recommends the discounting of liabilities at the accumulative rate. The use of the accumulative rate is not acceptable per se but the rationale of using a safe net of tax rate is sound.

2 Expenditure

This is optional and will only be undertaken if it provides a sufficient return. This 'sufficient return' is taken for convenience to be the rate of return that the investment as a whole provides. Thus expenditure is discounted at the remunerative rate.

The distinction may be of vital importance to investment decisions.

Example 9.6

B offers £1,000 to A for the leasehold interest. A receives a profit rent of £380 pa with 14 years remaining and no rent reviews on head- or sub-lease. A has agreed to carry out repairs at a cost of £1,000 in 4 years' time. Should B's offer be accepted?

Assuming a leasehold rate of 11% and 3% adj tax at 40%:

Profit rent	£380 pa	
YP for 14 years at 11% and 3% adj tax at 40%	4.8183	£1,830
Less: Cost of repairs		
(l) Liability?	£1,000	
PV £1 in 4 years at 3%	0.8884	£888
Estimated capital value (l)		£942
or :		
(2) Expenditure	£1,000	
PV £1 in 4 years at 11%	0.65873	£658
Estimated capital value (2)		£1,172

The distinction is obviously vital. If the cost is treated as a liability then the offer is accepted but not if treated as an expenditure. In many cases the decision is an arbitrary one, the effect of which can be reduced by the use of realistic 'safe' money rates.

A further problem now emerges and that is the estimation of future expenditures and liabilities when they are not known or fixed in monetary terms at the date of valuation. For example, an expected major renewal in four years' time can only initially be estimated on the basis of cost today. But the deduction to be made from market value in good condition must be an amount the market considers fairly reflects the current condition. Where the future sum is fixed in monetary terms then a present value calculation at a realistic monetary safe rate is satisfactory. Where it is a current estimate then the valuer must consider whether the costs will increase at a faster rate than that which can be safely earned on savings.

If the expenditure is likely to rise at 10% pa and money can be saved to earn interest at 10% pa, then the wisest solution is to deduct the full cost from today's value. If money can earn interest at a higher rate than the estimated inflationary increase in costs then a discounted sum can be deducted. If costs are rising faster than money rates then it would be logical to deduct the full cost now and indeed to have the repairs or renewals undertaken now. However, unless the work is essential as at the date of valuation, it may be as realistic to simply write down the value in sound condition by an appropriate factor.

Extensions and renewals of leases

The occupier or tenant of business premises will often be anxious to remain in occupation because a business move might involve considerable expense, loss of trade and loss of goodwill, resulting in a large loss of profit. In these circumstances tenants will be keen to negotiate an extension of the lease, a renewal of the existing lease on similar terms, or the grant of a completely new lease on different terms. Similarly landlords anxious not to lose a tenant may commence negotiations for a new lease two to three years prior to the end of a lease and ahead of the statutory or contractual date for service of notice to terminate the lease.

The problem of a tenant who approaches the landlord as the lease draws to an end provides very little difficulty. The landlord will require a rent approaching, or at, market rental value, and the tenant will expect to pay it, because that is what it would cost to lease a comparable property. But finding alternative accommodation is not an easy operation, and business decisions are prudently taken well in advance. It is thus more usual for a tenant to approach the landlord well before the termination of the existing lease and when they do, a valuation problem will arise. If the tenant approaches the landlord during the currency of the lease with a proposal to renew that lease immediately for an extended period, it will follow that the tenant is offering to surrender the current lease. (Such problems are often called 'surrender and renewals'.)

If, as is probably the case, a profit rent or a notional profit rent has arisen, then any surrender will be a surrender of valuable leasehold rights in the property. The tenant would be ill-advised to accept a new lease at market rental value.

The landlord, on the other hand, is not likely to agree to any indiscriminate extension of the tenant's profit rent because the anticipated reversion to market rental value will already be reflected in the market value of the freehold. Negotiations must be conducted to see that, following the surrender and lease renewal, there is no diminution in the value of the landlord's or of the tenant's interests in the property.

Valuers acting for both parties will be checking the position from both sides on the basis that the present interest should equal the proposed interests. This will involve four or more valuations.

Example 9.7

A tenant occupies promises on a 21-year lease with 2 years to run at £15,000 a year. The tenant wishes to surrender this lease for a new lease for 15 years with rent reviews every five years. The MRV on FRI terms for 15 years with 5 year reviews is £30,000. Freehold ARY is 7%.

Present interest Tenant		Present interest Freeholder		
MRV	£30,000		£15,000	
Rent payable	£15,000	YP 2 years at 7%	1.8080	
				£27,120
Profit rent	£15,000		£30,000	
YP 2 years at 8%		YP perp at 7%		
and 3% tax at 40%	1.1099	× PV £1 in 2 years at 7%	12.4777	
	£16,648			£374,331
				£401,451

Proposed interest Tenant			Proposed interest Freeholder		
MRV	£30,000			£x	
Rent payable	£x		YP 5 years @ 7%	4.1002	£4.1002x
	£30,000 – x			£30,000	

YP 5 years at 8% and 3% tax at 40%		2.538	YP perp at 7% × PV £1 in 5 years at 7%	10.1855	
		£76,157 – 2.5386x			£305,565
					£305,565 + 4.1002x

Equating present with proposed interests:

Tenant's position

£16,648	=	£76,158 – 2.538x
£59510	=	2.5388x
£23,442	=	x

Freeholder's position

£401,451	=	£305,565 + 4.1002x
£95886	=	4.1002x
£23,385	=	x

Leading to a negotiated settlement at, say, £23,400.

The landlord requires a minimum rent of £23,385, while the tenant can afford to offer £23,442. Negotiation between the parties will take place. A split at around £23,400 may be the result, or the parties may settle at a figure nearer the landlord's or tenant's figure depending upon their negotiating strength and the state of the market.

It must be pointed out that the above type of solution may result in a tenant's bid being lower than the landlord's minimum requirement. If the gap is sufficiently large, the proposals may be shelved. Considerable forces of inertia will, however, normally conspire to produce an agreement if the shortfall is of a minor nature: for example, the landlord, if satisfied with the tenant, may wish to save the advertising and legal fees involved in finding a new tenant. And the old tenant will have many reasons, already discussed, for being prepared to make a small loss in order to carry on in occupation of the premises.

Example 9.8

T occupies 6 High Street, holding a lease from L at a rent of £2,000 pa FRI with eight years remaining. T requires a new 40-year lease, starting immediately, and proposes to carry out improvements to the premises in 3 years' time at an estimated cost of £12,000 which will increase market rental value by £2,500 pa.

As a condition of the present lease, L requires that T pays a premium of £1,000 in 5 years' time. It is proposed that under the new lease T should pay a premium of £3,000 immediately and £5,000 after 20 years.

7 High Street is an identical property and has recently been sold on a 10% basis. It has just been let on a 21-year lease at a rent of £6,000 pa FRI with a premium of £15,000 payable at the start of the lease by the lessee.

It is agreed that the rent for 6 High Street should increase by 50% halfway through the proposed term.

Acting between the parties, assess what rent should be fixed under the proposals.

This problem is rather involved and includes examples of everything discussed in this chapter. It must be read and analysed extremely carefully before being attempted, and its implications must be fully realised. For example, if improvements costing £12,000 will increase MRV by £2,500 pa, this is a fact that cannot be ignored when assessing the value of the landlord's present interest, because it might be possible for the landlord to obtain possession in 8 years' time and carry out the said improvement.

It will be necessary to calculate the market rental value of 6 High Street from the information given concerning No 7.

MRV of 7 High Street = £6,000 + annual equivalent of £15,000 premium.

(£6,000 is a reduced rent to compensate for the premium. The market rental value will thus be £6,000 plus the value of the premium spread over the 21-year term.)

From the landlord's point of view =

$$£6,000 + \frac{£15,000}{\text{YP for 21 years at 10\%*}}$$

From the tenant's point of view =

$$£6,000 + \frac{£15,000}{\text{YP for 21 years at 11\% and 3\% adj tax at 40\%}}$$

*Property (freehold) sold on a 10% basis.

YP for 21 years at 10%	=	8.6487
YP for 21 years at 11% and 3% tax at 40%	=	5.9481
Average YP = (8.6487 + 5.9481)/2	=	7.2984
£15,000/7.2984	=	£2,005 pa
Therefore MRV = £8,055 FRI (#6,000 + #2,005)		

A Landlord's present interest

Rent	£2,000 pa
YP for 8 years at 9%	5.5348
	£11,070

Reversion (1): If landlord does not carry out the possible improvements.

Rent	£8,055 pa
YP rev perp after 8 years at 10%*	4.665
	£37,577

Reversion (2): Assuming landlord does carry out the improvements, and assuming the delay is short enough to involve no appreciable loss of rent.

Rent	£10,555 pa
YP rev perp after 8 years at 10%	4.665
	£49,239

*Both rents are full rental values, although they differ in magnitude, the security of income is assumed unchanged.

The figure of £49,239 can only be achieved by an expenditure of £12,000 in 8 years' time and therefore value today must be reduced accordingly.

Cost of improvements	£12,000
PV £1 in 8 years at 10%*	0.4665
	£5,600

*This is 'expenditure' and discounted (possibly erroneously) at the remunerative rate.

This leaves a net present value for the reversion of £43,639 (£49,239 – £5,600).

    Reversion (2) is more valuable so the landlord would improve, and the value of the freehold interest is this increased sum.

Estimated capital value	
£11,070 + £43,639 =	£54,709

In addition, a premium is payable after 5 years:

Premium		£1,000
PV £1 in 5 years at 10%	0.621	£621
Estimated capital value		£55,330

B Landlord's proposed interest

Rent		x	
YP for 20 years at 9%		9.1285	9.1285x
Reversion to rent		1.5x*	

*Rent increased by 50%

YP for 20 years at 9.5%	8.8124		
PV £1 in 20 years at 9.5%	0.1628	1.4348	2.1523x
Reversion to MRV		£10,555*	
YP rev perp after 40 years at 10%		0.221	£2,333
Plus premiums:	1 Immediately		£3,000
	2 In 20 years	£5,000	
	PV £1 in 20 years at 10%	0.14864	£743
			11.2808 x + £6,076

*Rent as increased by tenant's proposed improvements.

Present interest	=	proposed interest
£55,330	=	£6,076 + 11.2808x
£49,354	=	11.2808x
£49,354/11.2808	=	x
£4,375 pa	=	x

C Tenant's present interest

Rent received	£8,055 FRI	
Rent paid	£2,000 FRI	
Profit rent	£6,055	
YP for 8 years at 11% and 3% adj tax at 40%	3.3622	
CV		£20,358
Less premium	£1,000	
PV £1 in 5 years at 7%* 0.713		£713
Net CV		£19,645

* Liability: so use a realistic, low, safe, net accumulative rate. This rationale obviously raises a question mark about a dual rate YP using an accumulative rate of 3%; but to amend that would involve the reappraisal suggested in Chapter 7.

D Tenant's proposed interest

Rent received	£8,055
Rent paid	*x*
Profit rent	£8,055 − *x*
YP for 3* years at 11% and 3% adj tax at 40%	1.5403
CV	£12,407 − 1.5403*x*

* After 3 years T improves the premises and increases MRV.

Reversion to:	
Rent received	£10,555
Rent paid	*x*
Profit rent	£10,555 − *x*

YP for 17 years at 11% + 3% adj tax at 40%	5.3594	
PV £1 in 3 years @ 11%	× 0.73	3.9188
CV		£41,363 − 3.9188*x*

Reversion to:	
Rent received	£10,555
Rent paid	1.5*x*
Profit rent	£10,555 − 1.5*x*

YP for 20 years at 11% + 3% adj tax at 40%		5.8131
PV £1 in 20 years at 11%	0.124	0.7208
CV		£7,608 − 1.0812*x*

Adding the constituent elements of the valuation together:

$$
\begin{array}{rcl}
£12,407 & - & 1.5403x \\
£41,363 & - & 3.9188x \\
£7,608 & - & 1.0812x \\
\hline
\end{array}
$$

Total   £61,378 – 6.5403x

Less

1   Improvements

	£12,000	
PV £1 in 3 years at 11%	0.7312	£8,774

*Expenditure

2   Premium now   £3,000

3   Premium in 20 years

	£5,000	
PV £1 in 20 years at 7%*	0.2584	
		£1,292
		£13,066

*Liability

New total CV   £48,312 – 6.5403x

Present interest	=	Proposed interest
£19,645	=	£48,312 – 6.5403x
6.5403x	=	£28,667
x	=	£4,383 pa

Minimum L will accept:	£4,375 pa
Minimum T can offer:	£4,383 pa
Agreement between parties of (say):	£4,380 pa

Example 9.8 is as originally set out in the first edition and is included here in the same form to illustrate how the traditional or conventional valuer would approach this complex problem. It is obvious that the advice to landlord and tenant based on such calculations could be wrong. Provided the basis is supported by market evidence the valuers could argue that the figures are reasonable reflections of the current market. The reader's eyebrows should have risen over the last few pages and the red pen used frequently for marginal comment. The tutor need search no further for new problems; they are all here if the issues are reconsidered on an equated yield, modified cash flow approach, or on a full DCF basis.

The valuer must have regard to and take account of all the peculiarities of conventional valuation, the problems of future liabilities entered at current costs, the problems of long reversionary leases without rent reviews and the way in which a surrender and renewal can alter the market's perception of both the freehold and leasehold interests.

Question

An office building was let 30 years ago for 40 years without a rent review clause in the lease, at £10,000 a year. It is now estimated that the building would let in the market at £100,000 a year. Similar office buildings in the vicinity that are let at market rentals have been sold recently on an 8% basis.

Calculate:
(a) the rent to be paid on the surrender of this lease for a new 25-year lease with 5-year rent reviews
(b) the price the landlord should pay the tenant to secure a new 25-year lease to the tenant at MRV.

Marriage value

Where a property is split into multiple interests, either physically or legally, or both, each of the newly created interests will have a market value. The total of these values will not necessarily equate with the market value of the freehold in possession of the whole property. In such cases, an element of what is known as 'marriage value' might be shown to exist.

The following example will serve to demonstrate this.

Example 9.9

A is the freeholder of an office block, the market rental value of which is £28,000 pa on FRI terms.

Fourteen years ago A let the whole to B on a 40-year lease at a rent of £10,000 pa on FRI terms without a rent review.

B sub-let to C 6 years ago, at a rent of £18,000 pa. FRI for a term of 25 years without rent review.

B wishes to become the freeholder in possession of the office block. Advise him on the sum to be offered for the interests of A and C.

How much should each accept?

Valuation of current interests:
(Freehold rate 10%)

1 A's interest:*

Rent	£10,000	
YP for 26 years at 9%	9.929	
		£99,290
Reversion to FRV:	£28,000	
YP rev perp after 26 years at 10%	0.839	£23,492
Total		£122,782

*Traditional term and reversion making a 1% reduction in yield for the term's secuity.

2 B's interest:

Rent received:		£18,000	
Rent paid:		£10,000	
Profit rent:		£8,000	
YP for 19 years at 10% and 3%			
adj tax at 40%		6.0112	
			£48,090

Reversion to rent received			
		£28,000	
Rent paid		£10,000	
Profit rent		£18,000	
YP for 7 years at 11% and 3%		3.0533	
adj tax at 40%			
PV £1 in 19 years @ 11%	0.1377	0.4204	£7,567
Total			£55,657

3 C's interest

Rent received	£28,000	
Rent paid	£18,000	
Profit rent	£10,000	
YP for 19 years at 11% and 3%		
adj tax at 40%	5.6703	
Estimated value		£56,703

The total value of all interests at present is £122,782 + £55,657 + £56,703, a total of £235,142.

B wishes to become freeholder in possession. How much will this be worth now that the full rental value can be received immediately and in perpetuity.

MRV	£28,000
YP perp at 10%	10
	£280,000

Notice that the total value of all interests at present is only £235,142 so there will be a marriage value, created by the merger of interests, of £44,858. Generally the market value of a freehold in possession will be a more attractive investment than a freehold subject to a number of leases and sub-leases. The reality is that this could become an 8% or 9% investment boosting the marriage value by a further £31,000 (£28,000 at a YP of 9% in perpetuity).

How is this marriage created, and where does it arise?

Capital value is the product of two things: income and a capitalisation factor. It follows that the marriage value must arise from one or both of these.

1   *Income:* Does this change upon merger of the interests? In the case of the freehold in possession, the total income passing is £28,000. When A, B and C have interests in the property A receives

£10,000, B makes a profit rent of £8,000 and C makes a profit rent of £10,000. The total of rents and profit rents passing is therefore £28,000.

This holds for any year. The total of rents and profit rents will always be equal to market rental value, because what the freeholder loses by way of rent, someone else gains as profit rent.

2   *Capitalisation factor*: Does this change? The freehold in possession is valued by applying a single rate YP to the whole £28,000. But when the interests are split, the profit rents earned by B and C are valued by dual rate YPs adjusted for tax.

This leads to a lower capital value:

YP for 10 years at 10%	=	6.1446
YP for 10 years at 10% and 3% adj tax at 40%	=	4.0752

The effect of splitting the freehold in possession into three interests has been to reduce the total value of the block owing to the effect of valuing the leasehold interests dual rate. When leasehold and freehold interests are merged to create a freehold in possession, the total value is increased because the whole income is valued single rate. The scale of the marriage value may in part be due to the use of dual rate adjusted for tax. The use of dual rate unadjusted immediately changes the scale of marriage value with a YP of 5.3410. The question as to whether marriage value is fact or fiction can only be answered in the market place.

This increase in value created by the merger is known as marriage value.

Marriage value	£44,856
C	£56,703
B	£55,658
A	£122,783
Unencumbered freehold:	£280,000

The merger of interests will create additional value and so B can offer £224,342 for the interests of A and C. B's present interest is worth £55,658: the freehold in possession will be worth £280,000 and the gain £224,342 (£280,000 − £55,658).

But how much should B offer to A and to C?

What price should A and C ask for their interests?

B's first move will be to buy either A's or C's interest. The maximum B will be able to offer to A is the gain to be made from the transaction. If A's interest is purchased the freeholder will only be subject to C's underlease.

Value of B + A

Rent	£18,000	
YP for 19 years at 9%	8.9501	
		£161,101
Reversion to	£28,000	
YP rev perp after 19 years at 10%	1.635	
		£45,780
Total		£206,881

B's present interest is worth £55,658: the gain will be £206,881 − £55,658 = £151,223 and this is the maximum that can be offered to A. The minimum that A will accept will be the market value of £122,783. Assuming that A and B will employ valuers who will be aware of both figures agreement will be reached between these two boundaries.

The difference between these figures is: £151,223 − £122,783 = £28,440 and is the marriage value between A and B.

It can also be found in the following way.

A, before the transaction, had an interest worth £122,783; B had an interest worth £55,658. This totals £178,441. The value of the combined interest is £206,881; and the difference between these two figures is £28,440, the marriage value.

B's next step will be to acquire the interest of C. It must be noted that B now has an interest, as freeholder, worth £206,881. By acquiring C's interest, B becomes the freeholder in possession with an interest valued at £280,000. The gain that B stands to make is therefore:

£280,000 − £206,881 = £73,119

This is therefore the maximum that B could offer to C.

The minimum that C will accept will be the market value of £56,703. There will again be negotiation between these two figures. The difference between these two figures is £16,416, ie the marriage value between B and C.

This may also be obtained by summating the present interest of B and C (£206,881 and £56,702 = £263,584) and deducting this from B's new interest worth £280,000.

£280,000 − £263,584 = £16,416

Note that the total marriage value was found to be £44,856.

This is split between A and B and B and C:

Marriage value A + B:	£28,440
Marriage value B + C:	£16,416
Total marriage value:	£44,856

This term 'marriage value' usually refers to the above, the result of the merging of interests in the same property. It can also be used to describe the extra value created by a merger of two properties.

Example 9.10

B is the owner of a derelict house on a small site. On its own, the site is too small to be profitably developed, but could be used as a parking space. A is in a similar position next door. The area is zoned for shopping use and an indication has been given that planning permission would be given to construct a small shop covering both sites.

Assess the price that B could offer to A for the freehold interest in his property, and a reasonable sale price.

Value of B's present interest:

Rent for parking, say	£50.00 pa
YP perp at 15%	6.67
CV	£333.50

Value of proposed merged site:

This might be valued by using a 'residual valuation' (see Chapter 11) but market evidence indicates that similar sites have been selling for £3,000.

Maximum B can offer	=	£3,000 – £333.50
	=	£2,666.50
say		£2,650
Minimum A will accept		£333.50

A price will be reached by negotiation between these two figures. A marriage value of £2,333 (£3,000 – £667) has been created in this case, by a merger of sites.

Throughout this chapter, traditional approaches to valuation have been followed. Elsewhere, a number of these approaches have been questioned. It is therefore necessary to emphasise that it is the valuer's role to assess market value. If the valuer feels confident that he can substantiate the particular approach adopted by reference to the market, then marriage value in multiple interest investments may well exist; but the valuer must be certain of this market fact. He must beware that it is not based on false assumptions and fortuitous mathematics.

For example, an approach adopted in the USA for valuing leasehold investments in property could be loosely described as the 'difference in value' method. The logic used is that if a property has a rental value of £28,000 a year, and on that basis has a market value of £280,000 but due to the grant of certain leases is currently worth only £122,782, then the value of the leasehold investments is £280,000 less £122,782. On this basis marriage value does not exist.

This, then, is one extreme. At the other, consider the value of a very short leasehold interest. The value is likely to be low owing to the short term and the problem of dilapidations. The freehold interest would, however, reflect fully the loss of rental until review and in such a case genuine marriage value would exist.

It should be emphasised that the single rate valuation of leasehold interests (see Chapter 7) will, in the absence of any significant adjustment of capitalisation rates, greatly reduce any apparent marriage value, and may produce valuations which have greater affinity with market realities. So too can the use of a real value or DCF equated yield approach.

A discussion of marriage value could not be complete without some mention of 'break-up' values. This in essence is the opposite of marriage value and recognises that different investors have different needs, objectives, risk preferences and tax positions. As one might purchase a company as a whole and then sell off separately the component parts to realise a higher total value so one might buy a freehold interest in a property and through careful sub-division of title realise a higher total value. The idea is really no more complex than that of lotting a large estate. This again emphasises the need to couple a thorough knowledge of the discounting technique with a thorough knowledge of the property market. A particular issue here is the whole question of unitisation of single property investments. The market will dictate whether 1,000 units in a £10 million property will be worth £10,000 per unit or more.

The test is not: do the calculations suggest that marriage value exists?

But: does the market evidence prove that marriage value exists?

Market rental value, non-standard reviews and constant rent theory

How much rent should be paid under a lease with non-standard rent review periods so as to leave the two parties in the same financial position as if they had agreed on a standard term?

In times of rental increases, or, indeed, decreases, the rent review pattern will affect the rental value of a property, a concept so far regarded as being inflexible. It will be to the landlord's advantage in times of rental growth to insist upon regular rent reviews. Conversely, if he is burdened or presented with an arrangement whereby few, or no, rent reviews are provided for in the lease, then he may require a higher initial rent as compensation.

Two situations may arise where such compensation may be applicable. First, a new lease may be arranged without the prevalent rent review pattern. For example, in a market where three-yearly reviews are normally accepted, a new lease may be granted with seven-yearly reviews. In such a case the landlord might ask for a higher initial rent. Second, the problem might arise at a rent review where the period between reviews is the result of previous negotiations. For example, a 42-year lease granted 21 years ago with a single midterm review will present problems if three-yearly reviews are currently accepted, and again the landlord might have to ask for a higher reviewed rent as compensation.

Several methods have been devised to deal with this problem in a logical manner. One method was illustrated in a letter by Jack Rose to the *Estates Gazette*, 3 March 1979, at p824.

This method is designed to produce a single factor, k, which is applied to the full rental value on the usual pattern, to arrive at the compensatory rental value. This is formulated as:

$$k = \frac{(1 + r)^n - (1 + g)^n}{(1 + r)^n - 1} \times \frac{(1 + r)^t - 1}{(1 + r)^t - (1 + g)}$$

where
k = multiplier
r = equated yield
n = number of years to review in subject lease
g = annual rental growth expected and
t = number of years to review normally agreed.

Example 9.11

What rent should be fixed at a rent review for the remaining seven years of a 21 year lease of shop property, when rents are expected to rise at 12% pa and three-yearly reviews are prevalent in the market?

An equated yield of 15% based on the return provided by undated government stock with an adjustment for risk should be used, and the full rental value with three-yearly reviews would be £15,000 pa.

$$k = \frac{(1 + 0.15)^7 - (1 + 0.12)^7}{(1 + 0.15)^7 - 1} \times \frac{(1 + 0.15)^3 - 1}{(1 + 0.15)^3 - (1 + 0.12)^3}$$

$$= \frac{1.660 - 2.211}{1.66} \times \frac{0.521}{1.521 - 1.405}$$

$$= 0.2705 \times 4.44914 = 1.2149$$

Rent to be charged: £15,000 × 1.2149 = £18,224 pa

There is growing evidence in the property market that the rent review pattern in a lease has a considerable effect upon the rents required by landlords, and the above is an attempt to make such adjustments logically. This method, and the others suggested, are adaptations of discounted cash flow techniques to valuation problems. It must be concluded that the use of DCF in all valuations will remove all of the inconsistencies referred to in this chapter as well as those problems identified in Chapters 5, 6 and 7, and is the ultimate and logical refinement of 'the income approach'.

Question

Rents on five-year rent review patterns are £25,000 a year. What rent should a landlord expect if the property is to be let with rent reviews every three years. Use a growth rate of 5% an equated yield of 10%.

The logic behind the DCF approach is difficult to defeat but the market is reluctant to move into the uncharted area of market forecasting. The valuer is trying to establish on the basis of the market rental value for a specific fixed term what the rent should be for a longer or shorter fixed term. The first is assumed to equate with the actual annual rentals each year for the term discounted to their present worth and re-assessed as a fixed annual equivalent sum. The second is the projection or shortening of the actual expected annual sums re-expressed as an annual equivalent. This process can only be undertaken on the basis of an assumed (but hopefully research-supported growth rate) or on the basis of an implied growth rate (see Chapter 13).

The latter approach is more questionable in this exercise because the implied rental growth figures are long-term averages to be used for the purpose of valuing non-market rent review patterns. Either the landlord or the tenant would be concerned if the actual rental value rose at a higher or lower rate than the implied average growth rate. For an uplift rent the valuer is trying to determine the appropriate growth rate for that specific property. There is then, an argument in favour of using forecasted growth rates based on thorough research for assessing uplift rents rather than merely using implied rates.

In arbitration cases it is a question of presenting the strongest supportable case for or against the uplift, the arithmetical assessment is only a starting point.

In the case of non-normal rent reviews the courts have adopted a simple 10–15% increase for a longer review or increased the market rental by a factor for each additional year over the normal rent review evidence in preference to a DCF approach.

The examples in this chapter do not reflect the statutory rights of landlords and tenants which may have to be taken into account and are considered in Chapter 10.

Upward/downward issues

The property industries Code of Practice for Commercial Leases in England and Wales, a self regulatory code prepared in response to concerns raised by government relating to the restrictive practice of upward only rent review and other lease arrangements, indicated in Recommendation 5 that 'Where alternative lease terms are offered, different rents should be appropriately priced for each set of terms'. How to assess those different rents led to a series of research exercises undertaken by valuation practitioners and academics between 1999 and 2003. In the interim the typical lease length has shortened, break clauses have been introduced into more leases, new comprehensive service provision included in multi-tenanted buildings and incentives offered as part of lease packages. All of this has added to the difficulty of rental analysis and rental comparison re-emphasising the need for considerable care in quoting rent per m². The issue is rent per m² of what space and on what terms.

Nevertheless the researchers were able to establish as a general principle that a landlord would require rental compensation where the rent review was to move from upward only to upward/downward. The amount of 'compensation' in the absence of sufficient comparable rental evidence would depend upon the markets expectations as to 'rental volatility', 'probability of tenant vacating' and 'expected costs of tenant vacation'. Simulation exercises have been developed and are used in practice to assess the rental adjustment needed to balance expected loss in income or capital value associated with a move from say, a 15 year lease with upwward only rent reviews every five years to, a ten year lease with upward/downward rent reviews. These simulation figures may be countered by the pragmatic view of tenants that they are unwilling to pay significant increases in rent for a more flexible lease arrangement.

These issues are important for advisors on rental terms and conditions and the implications of upward/downward rent reviews will need to be reflected by valuers when valuing properties subject to such lease terms. If the rental adjustment is correct and fair there may be no impact on capitalisation rates but where there are flexible lease terms without proper rental adjustment there could be an impact on capitalistion rates.

For further information readers are referred to Baum, A (2003) *Pricing the options inherent in leased property: a UK case study*, ERES conference, Helsinki, 15 June 2003.

Summary

- Premiums are payable by tenants to landlords to ensure a letting at below market rental or for other reasons.
- Reverse premiums are payable by landlords to tenants or by tenants to sub-tenants to secure deals above market rentals.
- Premiums are generally assessed on the basis of value equations. Both parties should be neither better off nor worse off, and a before and after valuation is preferred.
- At the end of a lease either landlord or tenant may seek to secure a new lease in exchange for the surrender of the old lease. Neither party should be any better or worse off following the surrender and renewal, and a present v proposed interest calculation is normally used.
- Where multiple interests exist in a single property marriage value may exist.
- Constant rent theory argues that a landlord should be no worse off over time by letting on an extended rent review pattern, but the market places an upper limit of 10%–15% on the uplift negotiable for longer intervals between rent reviews.
- A similar issue has emerged with the move from upward only rent reviews to upward/downward rent reviews. Here again one can establish through the use of DCF techniques and simulation the amount of additional rent a landlord would require to compensate for the loss of security attaching to upward only rent reviews but tenants may not be willing to meet such uplifts from year one. Tenants would argue that if rents do rise then they may well end up paying twice, once through the uplift and once through the increased rent at the rent review.

The Effects of Legislation 10

In the examples shown in the text it has been assumed that landlords and tenants are generally free to negotiate whatever lease terms they find acceptable for a particular tenancy of a property. In particular it has been assumed that they are free to agree the amount of rent to be charged for the premises and that in most cases the landlord can obtain vacant possession at the termination of the current lease. This is not always the case in the UK and the valuer must have regard to the provisions of the Landlord and Tenant Acts and the various Rent and Housing Acts as they affect properties occupied for business and residential purposes. The Agricultural Holdings Acts are of similar importance in the case of agricultural property, but agricultural property is not considered in this text.

The predominant feature of all landlord and tenant legislation is that it amends the normal law of contract as between landlord and tenant giving tenants substantial security of tenure (the right to remain in occupation following the termination of a contractual agreement) and setting out statutory procedures for determining and/or controlling the rents that a landlord can charge a tenant for the right to occupy a property. (The subject-matter of this chapter is discussed in more detail in Statutory Valuations.)

Business premises

The most important statutes affecting business premises are the Landlord and Tenant Act 1927 Part I and the Landlord and Tenant Act 1954 Part II as amended by the Law of Property Act 1969 Part I and by the Regulatory Reform (Business Tenancies) (England & Wales) Order 2003.

These statutes primarily affect industrial and commercial property, but the expression 'business' includes a trade, profession or employment and includes any activity carried on by a body of persons whether corporate or not incorporate. This definition as set out in section 32, Landlord and Tenant Act 1954 Part II is sufficiently broad to include some types of occupation which would not normally be regarded as business occupations, such as a tennis club (see *Addiscombe Garden Estates Ltd* v *Crabbe* [1958] 1 QB 513).

Compensation for improvements

Under the Landlord and Tenant Act 1927 Part II the tenant of business premises is entitled 'at the termination of the tenancy, on quitting his holding, to be paid by his landlord compensation in respect of any improvement (including the erection of any building) on his holding made by him or his predecessors in title, not being a trade or other fixture which the tenant is by law entitled to remove, which at the termination of the tenancy adds to the letting value of the holding' (section1, Landlord and Tenant Act 1927).

This right does not extend to improvements carried out before 25 March 1928, nor to improvements 'made in pursuance of a statutory obligation, nor to improvements which the tenant or his predecessors in title were under an obligation to make, such as would be the case where a tenant covenanted to carry out improvements as a condition of the lease when entered into' (section 2, Landlord and Tenant Act 1927). Except that those made after the passing of the 1954 Act 'in pursuance of a statutory obligation' will qualify for compensation (section 48, Landlord and Tenant Act 1954).

The tenant will normally require the consent of the landlord before carrying out alterations or improvements, or alternatively they may apply to the court for a certificate to the effect that the improvement is a 'proper improvement' (section 3, Landlord and Tenant Act 1927).

It should be noted under section 19(2), Landlord and Tenant Act 1927 Part II that

> in all leases ... containing a covenant, condition or agreement, against the making of improvements without licence or consent, such covenant ... .shall be deemed ... to be subject to a proviso that such licence or consent is not to be unreasonably with-held ...

Section 49 of the 1954 Act renders void any agreement to contract out of the 1927 Act.

Compensation payable is limited under schedule 1 of the Landlord and Tenant Act 1927 to the lesser of:

1. the net addition to the value of the holding as a whole as a result of the improvement or
2. the reasonable cost of carrying out the improvement at the termination of the tenancy, subject to a deduction of an amount equal to the cost (if any) of putting the works constituting the improvement into a reasonable state of repair, except as so far as such cost is covered by the tenant's repairing covenant.

Further, 'in determining the amount of such net addition regard shall be had to the purposes for which it is intended the premises shall be used after the termination of the tenancy' (section 1(2), Landlord and Tenant Act 1927). For example, if the premises are to be demolished immediately then the improvements are of no value to the landlord and no compensation would be payable. But if the premises are to be demolished in, say, six months' time and there is a temporary user planned then compensation would be based on the net addition to value of the improvements for that six month period. If the landlord and tenant fail to agree as to the amount of compensation, the matter can be referred to the county court. Where a new lease is granted on the termination of the current lease no compensation can be claimed at that point in time. Both the 1927 Act and the 1954 Act provide that the rent on a new lease shall exclude any amount attributable to the improvements in respect of which compensation would have been payable (see '(c)', p178).

Thus the initial problem to be solved by a valuer when instructed to value business premises is the extent to which they have been improved by the tenant and the extent to which these may become compensatable improvements under the provisions of the Landlord and Tenant Act 1927.

Example 10.1

Value the freehold interest in office premises currently let at £50,000 on full repairing and insuring terms with 6 years of the lease unexpired. The current market rental value on full repairing and insuring terms is £100,000. Improvements were carried out by the tenant and these have increased the market rental value by £10,000 and would cost today an estimated £80,000 to complete.

Value the premises on the assumption that the tenant will vacate on the termination of the present lease and will be able to make a valid claim for compensation under the Landlord and Tenant Act 1927.

Current net income	£50,000	
YP for 6 years at 8%	4.62	£231,000
Reversion in 6 years to	£100,000	
YP perp deferred 6 years at 8%	7.87	£787,000
		£1,018,000

Less compensation under Landlord and Tenant Act 1927 for improvements being the lesser of:

(a) Net addition to the value

Increase in rental attributable to improvements:

	£10,000	
YP perp at 8%	12.5	
	£125,000	

(b) Cost of carrying out improvements

	£80,000	
Therefore compensation is	£80,000	
× PV £1 for 6 years at 8%	0.63	
	£50,400	

(b) is the lesser amount
Therefore
Value of freehold allowing for payment of compensation to tenant = £1,018,000 – £50,400 = £967,600
say £967,750

As shown elsewhere this traditional equivalent yield valuation raises some basic issues: first is the whole question of implicit versus explicit growth valuation models, the second is the validity of using current estimates of changes in rental value attributable to the improvements and of using current costs for assessing the compensations when the actual compensation will have to be based on figures applicable in six years' time. This is a further argument for at least checking the valuation by reference to a true DCF approach.

It could be argued that the £80,000 should not be discounted as in effect it is the equivalent cost of the future amount discounted. In which case a prudent purchaser would only offer £938,000 (£1,018,000 – £80,000).

While valuers may be asked to assess the amount of compensation to be paid under a 1927 claim it is rare for market values to be affected by the future possibility of such payments. Such a valuation is only likely to occur when it is known that a tenant who has undertaken improvements will be vacating at a specific date in the future.

Security of tenure

In accordance with the provisions of section 24(1) of the Landlord and Tenant Act 1954 tenants of business premises are granted security of tenure; however, the parties may contract out of these provisions. Contracting out requires the landlord to serve notice on the tenant or prospective tenant in a prescribed form under schedule 1 of the Regulatory Reform (Business Tenancies) (England and Wales) Order 2003 (RRO/03). Tenancies to which this part of the Act applies will not come to an end unless terminated in accordance with the provisions of Part II of the Act, so that some positive act by the landlord or tenant needs to be taken to terminate a tenancy. Where notice to terminate is served by the landlord he must give at least six months' notice and not more than 12 months' notice. Such notice cannot come into force before the expiration of an existing contractual tenancy. Thus in the case of most leases of business premises the earliest date a landlord can serve notice on a tenant is 12 months prior to the contractual termination date. If notice is not served the tenancy continues as a statutory tenancy at the contracted rent until terminated by notice.

The Act further provides that a tenant has the right to the renewal of his lease. If the landlord wishes to obtain possession he may oppose the tenant's request for a new tenancy only on the grounds set out in the Act.

Section 30(1) states the following grounds on which a landlord may oppose an application:

(a) where under the current tenancy the tenant has any obligations as respects the repair and maintenance of the holding that the tenant ought not to be granted a new tenancy in view of the state of repair of the holding, being a state resulting from the tenant's failure to comply with the said obligations

(b) that the tenant ought not to be granted a new tenancy in view of his persistent delay in paying rent which has become due

(c) that the tenant ought not to be granted a new tenancy in view of other substantial breaches by him of his obligations under the current tenancy, or for any other reason connected with the tenant's use or management of the holding

(d) that the landlord has offered and is willing to provide or secure the provision of alternative accommodation for the tenant, that the terms on which the alternative accommodation is available are reasonable having regard to the terms of the current tenancy and to all other relevant circumstances, and that the accommodation at the time at which it will be available is suitable for the tenant's requirements, including the requirement to preserve goodwill, having regard to the nature and class of his business and to the situation and extent of and facilities afforded by the holding

(e) where the current tenancy was created by the sub-letting of part only of the property comprised in a superior tenancy and the landlord is the owner of an interest in reversion expectant on the termination of that superior tenancy, that the estimate of the rents reasonably obtainable on separate lettings of the holding and the remainder of that property would be substantially less than the rent reasonably obtainable on a letting of that property as a whole, that on the termination of the current tenancy the landlord requires possession of the holding for the purpose of letting or otherwise disposing of the said property as a whole, and that in view thereof the tenant ought not to be granted a new tenancy

(f) that on the termination of the current tenancy the landlord intends to demolish or reconstruct the premises comprising the holding or a substantial part of those premises or to carry out substantial work of construction on the holding or part thereof, and that he could not reasonably do so without obtaining possession of the holding

(g) subject as hereinafter provided, that on the termination of the current tenancy the landlord intends to occupy the holding for the purposes, or partly for the purposes, of a business to be carried on by him therein, or as his residence.

To oppose an application under that last mentioned ground (g) the landlord must have been the owner of the said interest for at least five years prior to the termination of the current tenancy. Section 6 of the Law of Property Act 1969 extends section 31(g) of the 1954 Act to companies controlled by the landlord and section 7 of the Law of Property Act 1969 has altered the effects of 31(f) of the 1954 Act so that a landlord wishing to oppose the grant of a new tenancy under that ground must now not only prove his intent to carry out substantial works of alteration but also that it is necessary to obtain possession in order to complete such works. Thus if a landlord can demolish and rebuild without obtaining possession and if the tenant is agreeable or willing to co-operate then the courts will allow the tenant to remain in possession.

Section 32 of the Landlord and Tenant Act 1954, while requiring the new lease to be in respect of the whole of the building, has now been amended by section 7 of the Law of Property Act 1969, which adds sections 31(A) and 32(1)(A) to allow a court to grant a new tenancy in respect of part of the original holding where the tenant is in agreement.

Compensation for loss of security

When a landlord is successful in obtaining possession the tenant may be entitled to compensation under section 37 of the Landlord and Tenant Act 1954.

> Where the Court is precluded ... from making an order for the grant of a new tenancy by reason of any of the grounds specified in paragraphs (e), (f) and (g) ... the tenant shall be entitled on quitting the holding to recover from the landlord by way of compensation an amount determined in accordance with the following provisions of this section ...

The amount of compensation payable will be two times the rateable value of the holding if for the whole of the 14 years immediately preceding the termination of the tenancy the premises have been occupied for the purposes of a business carried on by the occupier or if during those 14 years there had been a change in the occupation and the current occupier was the successor to the business carried on by his predecessor. In all other cases the amount of compensation shall be the rateable value of the holding.

Additional compensation is now payable under section 37A. This is payable if a landlord obtains possession or is successful in opposing a tenant's request for a new lease and is shown subsequently to have succeeded due to misrepresentation or has concealed material facts. The amount in theses cases is determined by the court as 'such sum as appears sufficient as compensation for damage or loss sustained by the tenant ...'. Assessment of such a claim is likely to be based on factors other than property factors such as loss of trade and profits.

Terms of the new lease

Where a new lease is granted then the new rent payable will normally be in accordance with the provisions of section 34 of the Act, particularly when the parties are in disagreement and the matter is referred to the county court for settlement.

The Law of Property Act 1969 has amended section 34 so that:

the rent payable under a tenancy granted by order of the Court under this part of this act shall be such as may be agreed between the landlord and tenant or as, in default of such agreement, may be determined by the court to be that at which, having regard to the terms of the tenancy (other than those relating to rent), the holding might reasonably be expected to be let in the open market by a willing lessor, there being disregarded:

(a)   any effect on rent of the fact that the tenant has or his predecessors in title have been in occupation of the holding;

(b)   any goodwill attached to the holding by reason of the carrying on thereat of the business of the tenant (whether by him or by a predecessor of his in that business);

(c)   any effect on rent of any improvement carried out by the tenant or predecessor in title of his otherwise than in pursuance of an obligation to his immediate landlord (see LPA amendment below);

(d)   in the case of a holding comprising licensed premises any addition to its value attributable to the licence, if it appears to the court that having regard to the terms of the current tenancy and any other relevant circumstances the benefit of the licence belongs to the tenant.

Items (a), (b) and (c) are those that valuers have most frequently to reflect in valuations of business premises.

Items (a) and (b) cause particular difficulty in assessment for, while it is simple to explain the meaning of these requirements, it is often extremely difficult to assess them in practice. Under item (a) the valuer must demonstrate that, for example, if the occupying tenant would bid £55,000 but the premises have a rental value as defined in the Act of only £50,000, then only £50,000 is payable. The potential tenant overbid of £5,000 in a free market must be disregarded.

Similarly under item (b) if it can be demonstrated that the premises are worth £50,000 but to any other tenant carrying on the same business are worth £55,000 then the £5,000 of business goodwill must be disregarded.

Section 1 of the Law of Property Act has extended the meaning of section 34 (c) to include tenants' improvements carried out at any time within 21 years of the renewal of the tenancy.

All the other terms and conditions of a new tenancy shall be as agreed between the parties, but if the parties cannot agree then sections 33 and 34 of the Landlord and Tenant Act 1954 require the court to restrict the terms of the tenancy to 15 years (see RRO/03) with appropriate rent reviews (Law of Property Act 1969 section 2 which adds subsection 3 to section 34 of the 1954 Act).

A number of Law Commission recommendations for amendment to the 1954 Act were implemented by the Regulatory Reform (Business Tenancies) (England and Wales) Order 2003 the most important of which from a valuation perspective was the move from a maximum new lease length of 14 years to 15 years. The latter being more in line with market practice linked with rent reviews every five years.

Example 10.2

Assuming the facts as stated in example 10.1, value the freehold interest and assume the tenant is granted a new 15-year lease with rent reviews in the fifth and tenth year and that the improvements were completed 3 years ago. The first step to resolve this problem requires the sorting out of the income flow. In six years time the lease is due for renewal, at a rental ignoring the effect on rent of the improvements (Landlord and Tenant Act 1954).

But there is some doubt as to the rent that could be charged at the rent reviews after 5 and 10 years. According to the Law of Property Act it would seem that no account should be taken of the value of the improvements for a period of 21 years.

If this argument applies then the rent on the reviews must once more be at a figure excluding any value attributable to the improvements provided a section 34 disregard or equivalent is included in the rent review clause. This reasoning, coupled with the specific provisions in the Landlord and Tenant Act 1954, would effectively result in no account being taken of the improvements at any time during the whole of the new lease.

The income position for valuation purposes is as scheduled below:

Time	Activity	Rent
−3 years	Completion of improvements	
Today's date	Date of valuation	£50,000
+6 years	Lease renewal disregarding effect of improvements	£90,000
+11 years	First 5 year review	£90,000
+16 years	Second 5 year review	£90,000
+18 years	Improvements 21 years old	
+21 years	Next lease renewal	£100,000

Solution, assuming cash flow as shown above (and see notes following the valuation)

Current net income		£50,000	
YP for 6 years at 8%		4.62	£231,000
First reversion to section 34 rent		£90,000	
YP for 15 years at 8%	8.5595		
PV £1 in 6 years at 8%	0.63017	5.3939	£485,455
Second reversion to MRV		£100,000	
YP perp def'd 21 years at 8%		2.4832	£248,320
			£964,775

Notes
1. No specific allowance has been made for compensation for improvements as the valuation assumes further renewals under Landlord and Tenant Act to the current tenant or successor in title.
2. Conventional equivalent yield valuation assumptions underlie the valuation, the review pattern is such that an equated yield or DCF explicit valuation should be used as a check.
3. The valuer can only make reasonable assumptions. To assume for valuation purposes that the tenant will forget they have rights under the 1954 Act would be inferring a definition of market value that excluded acting prudently and with knowledge. To assume that the tenant's solicitors will forget to include a section 34 disregard in the rent review clause would be a false assumption and could give rise to an action for negligence against the valuer. The assumptions made in this solution are realistic and reasonable.

If a 21-year period elapses prior to a rent review date then the market rental value of the improved premises could be charged from the review date.

This would appear to be contrary to the 1954 Act as originally drafted which clearly intended that the effects on rent of any improvements should be disregarded for the whole of the new tenancy in this case for the whole 15 years.

While some confusion apparently exists, a number of points are becoming obvious from the decisions reached in a number of landlord and tenant cases.

First, the legislation relates quite clearly to the determination of 'rent payable under a tenancy granted by order of the Court', and this can only occur on renewal of a lease and not on a rent review.

Second, the Act uses the word 'reasonable' which suggests that the rent as determined need not be the maximum possible rent.

Third, the rent on any review will be determined in accordance with the appropriate clauses in the lease and, unless there is specific reference to section 34 of the Landlord and Tenant Act 1954 or a specific statement that improvements carried out during an immediately preceding lease, or within 21 years of the review date, are to be disregarded, the review rent may fully reflect the current rental of the property as improved.

Professional advisers are therefore forewarned when acting for tenants to see that rent review clauses in leases are sufficiently worded to protect their clients. In this example if the tenant was to accept as the terms of the new lease a rent review clause which contained no disregards then the market rent would be charged at the first rent review.

When licences for improvements are granted they should also confirm that the effect on rental value of the improvements will be disregarded on review or renewal of the current lease.

If this advice is followed then tenants may avoid any repetition of *Ponsford* v *HMS Aerosols Ltd* [1977] 2 EGLR 68 where tenants of a factory rebuilt the premises which had been destroyed by fire, at the same time substantially improving the property with the landlord's consent. Shortly after rebuilding, the rent was due for review.

The wording of the lease and licence was such that the Court of Appeal held that the tenants would have to pay rent in respect of the improvements, the cost of which they had borne themselves. The wording of the lease, together with the wording of any licences, will determine the factors to be taken into account when assessing the rent to be paid on review. A similar situation would occur if a tenant were to carry out improvements without the proper licence of the landlord. (Readers are also referred to *English Exporters (London) Ltd* v *Eldonwall Ltd* [1973] Ch 415.)

When valuing business premises, or advising tenants of business premises, the valuer must have regard to all the terms and conditions of the lease and to the relevant statutory provisions.

In practice, negotiations and/or court proceedings may result in the new rent commencing many months after the termination of a current contractual lease. Section 64 of the 1954 Act further provides that any new terms, including those relating to rent, may only commence three months after the court application has been 'finally disposed of'. This can lead to a considerable loss of income for the landlord.

The Law of Property Act 1969 added a new section, 24A, to the 1954 Act.

(1) The landlord of a tenancy to which this part of this Act applies may:

   (a) if he has given notice under section 25 of this Act to terminate the tenancy; or
   (b) if the tenant has made a request for a new tenancy in accordance with Section 26 of this Act; apply to the court to determine a rent which it would be reasonable for the tenant to pay while the tenancy continues by virtue of Section 24 of this Act, and the court may determine a rent accordingly.

(2) A rent determined in proceedings under this section shall be deemed to be the rent payable under the tenancy from the date on which the proceedings were commenced or the date specified in the landlord's notice or the tenant's request, whichever is the later.

The Regulatory Reform Order 2003 extended this right to tenants. Landlords are likely to exercise this right when rents have risen and tenants when rents have fallen. The interim rent will normally be the same as the rent determined for the new tenancy which means that, for normal valuation purposes, a valuer can assume that the new rent at reversion begins at the date of contractual termination of the old lease rent. The real position from a property management perspective may be different and this rental position will normally only be achieved if the right notices are served at the right times in accordance with the legislation as amended by the order. There may still be a case for interim rents to be settled on the basis of a tenancy from year to year. This 'cushioned' rent is only likely to occur where a landlord has applied for an interim rent and has opposed the grant of a new tenancy. A rent from year to year tends to be lower than a normal market rent.

Landlord and tenant negotiations

It is important to appreciate that whilst the Landlord and Tenant Acts are there to protect the tenant on termination of his lease, many tenants will seek to negotiate new leases before their current leases expire. The Landlord and Tenant Acts give the tenant increased bargaining strength and full regard should be given to these statutes when advising a landlord or tenant on the terms for a new lease.

Example 10.3

A tenant occupies shop premises on a lease having two years to run at £60,000 a year net. Ten years ago she substantially improved the property. The market rental value of the property today as originally demised would be £100,000 but as improved it is worth £140,000. The tenant wishes to surrender her present lease for a new 15-year lease with reviews every 5 years to MRV; the landlord has agreed in principle and you have been appointed as independent valuer to assess a reasonable rent for the new lease. Freehold capitalisation rates are 7%.

As outlined in Chapter 9, this requires consideration from the tenant's and landlord's points of view on a 'before and after' basis.

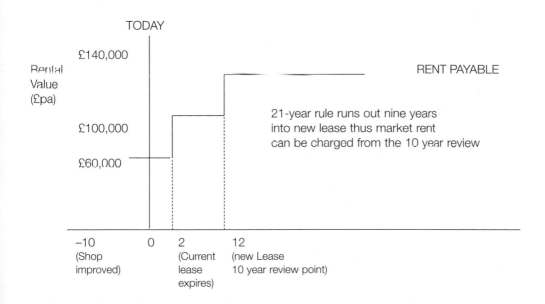

Note: The 21-year rule will run out after 9 years of the new lease and MRV can be charged from the rent review in year 10; that is, in 12 years' time.

Value of freeholder's present interest if she does not accept surrender of the lease:

Current income		£60,000	
YP for 2 years at 7%		1.8080	£108,480
Reversion to MRV subject to section 34(c) L&T Act 1954 Part II and assuming a new 15-year lease with review agreed		£100,000	
YP for 10 years at 7%	7.02		
PV £1 in 2 years at 7%	0.87	6.10	£610,000
Reversion to MRV		£140,000	
YP perp def'd 12 years at 7%		6.34	£887,600
Value of present interest			£1,606,080

(If the assumption here is that a review after 5 years would be permitted by the courts then one must further assume that professional advisers would see that the rent review clause fully reflected the intention of the Landlord and Tenant Act 1954 Part II as amended by the Law of Property Act 1969.)

Value of freeholder's proposed interest:

Let rent to be reserved for new 15-year lease $= £x$ pa		
Proposed rent	$£x$	
YP 5 years at 7%	4.10	$4.10x$
Reversions to MRV	£140,000	
YP perp def'd 5 years at 7%	10.185	£1,425,900
Value of proposed interest :	$£1,425,900 + 4.10x$	

On the assumption that the freeholder should be no better off and no worse off the value of his present interest must be equated with the value of his future interest:

Present interest	$=$	proposed interest
£1,606,080	$=$	$£1,425,900 + 4.1x$
£1,606,080 $-$ £1,425,900	$=$	$4.1x$
£180,180	$=$	$4.1x$
£43,940	$=$	x

Value of tenant's present interest assuming no surrender:

MRV to tenant		£140,000	
Less rent reserved		£60,000	
Profit rent		£80,000	
YP for 2 years at 10%		1.7355	£138,840
Reversion to		£140,000	
Less section 34 rent reserved		£100,000	
		£40,000	
YP for 10 years at 10%	6.1446		
PV £1 in 2 years at 10%	0.82645	5.078	£203,128
Value of present interest			£341,968

Note: After the rent review in 10 years the rent rises to £140,000

Value of tenant's proposed interest:
Let rent to be reserved for new 15-year lease = £x pa, then new profit rent is:

MRV to tenant	£140,000
Less rent reserved	x
Profit rent	£140,000 − x
YP for 5 years at 10%	3.7908
Value of proposed interest	£530,712 − 3.7908x

Present interest	=	proposed interest
£341,968	=	£530,712 − 3.7908x
3.7908x	=	£530,712 − £341,968
3.7908x	=	£188,744
x	=	£49,790

Here it would seem reasonable for the parties to accept a rent for a new 15-year lease, on surrender of the present two-year term, of, say, £45,000 a year with a market rent review after five years. But these valuations should be cross-checked with a DCF approach. A similar calculation using dual rate of 8% adjusted for tax at 40% would produce a similar conclusion although the leasehold figures would be different.

Question

Your client owns the freehold interest in an office building. The building has a ground floor and four upper floors and the whole is let on full repairing and insuring terms for a term of 21 years without reviews. The lease has three years to run at £150,000 a year.

The building originally contained only a ground and two upper floors but the tenant agreed as a condition of the lease to add a third floor. This work was completed within two years of the grant at a cost of £150,000. Seven years ago a mansard roof was added at a further cost of £50,000. All these works were approved by the landlord.

There are 400m² on each of the ground, first, second and third floors but the top floor has an area of only 250m².

Office space is letting at £200 per m². Where there is no lift, the rent on upper floors is £150 per m².

The property is assessed for rating at RV £350,000.

Freehold equivalent yields are 7%.

The tenants now wish to refit and equip the building with new carpets, lighting, computer, etc. but before proceeding wish to improve their security of tenure. They would prefer to buy the freehold but as a second best would like to surrender their present lease for a new 20-year full repairing and insuring lease with reviews every five years.

Advise the freeholder, (ignoring VAT):

(a) on the rent that could be expected in three years' time
(b) on the price that should be asked for the freehold interest
(c) on the rent that should be asked for the proposed 20-year lease
(d) on the amount of compensation due to the tenant if possession was recovered.

Residential property

In this section the relevant Acts will be referred to by their initials and their year. This is due to the variety of these, and the complexity of interaction between them. Thus Housing Acts will be HA 19yy, Rent Acts as RA 19yy, Landlord and Tenant Acts as LTA 19yy. Other Acts will be introduced for the first time, and then follow the same nomenclature.

Over 68% (2001 Census) of homes in England and Wales are now owner-occupied and the appropriate method for the valuation of such freehold residential property is that of direct capital comparison (Mackmin: *Valuation and Sale of Residential Property*, Routledge 1994). However, there are still many individuals and families occupying tenanted property both on a furnished and an unfurnished basis. The capital value of such properties may be assessed either by using the income approach or capital comparison. If capital comparison, then by direct reference to a vacant possession (VP) value, or a deferred VP value, or a percentage of the VP value, dependant upon the analysis of the comparables.

Since 1980 there have been significant changes in political attitude towards the public and private sectors of the rented residential market. A large number of public sector tenants exercised their 'right to buy' under the HA 1980 (subsequently HA 1985) and are now home owners. HA1988 made further significant changes to the law relating to public and private sector housing, including Housing Associations. As from 15 January 1989 all existing and new assured tenancies come under the HA 1988 (amended by HA 1996). HA 1988 also introduced assured shorthold tenancies (ASTs) to replace the previous system of protected shorthold tenancies. The Office of the Deputy Prime Minister (now the Department for Communities and Local Government) has a series of useful papers on the current situation on ASTs, 'right to buy' and discounts, etc that are normally kept up-to-date on its web site at *www.odpm.gov.uk* or *www.dclg.gov.uk*. If the valuer is dealing with property subject to older tenancies (especially those created before 15 January 1989) additional advice may be required.

The HA 2004 substantially increases local authority control in assessing housing conditions and enforcing housing standards, licensing 'houses in multiple occupation' (HMOs) and other residential accommodation, as well as introducing home information packs (HIPs) prior to marketing for sale and making various changes to 'right to buy' qualification periods, discounts and discount pay-backs. At the time of writing there is also a proposal to provide local authorities with powers to apply licensing to the private rented sector in part or all of their area (selective licensing), in order to address issues of poor management and anti-social behaviour by tenants. When valuing residential property the valuer will need to ensure that any such improvement notice, prohibition order or hazard awareness notice (Part 1 HA 2004), HMO licence (Part 2 HA 2004), selective licensing (Part 3 HA 2004) or other orders and notices under Parts 1 and 4 HA 2004 served by the local authority are considered.

The Leasehold Reform Act 1967 [LRA 1967] introduced the right to enfranchise (or extend the lease) of a house. The Leasehold Reform Housing and Urban Development Act 1993 [LRHUDA 1993] extended this concept to flats. The Commonhold and Leasehold Reform Act 2002 [CLRA 2002] has further amended the provisions of both LRA 1967 and LRHUDA 1993, as well as introducing the new tenure system 'commonhold'.

Private sector tenancies

Residential investment properties in the private sector can be grouped under the following heads:

1. assured tenancies under the HA 1988, as amended HA 1996
2. assured shorthold tenancies under the HA 1988, as amended HA 1996.

The following residential investment properties are excluded from being either assured or assured shorthold tenancies:

- Tenancies subject to protection under Part I RA 1977, generally referred to as regulated furnished or regulated unfurnished tenancies, including those tenancies previously known as controlled tenancies by section 64 HA 1980, except those to which Part II LTA 1954 applies. Tenancies in this category must have been 'entered into before, or pursuant to a contract made before, the commencement of the HA 1988', ie 15 January 1989.
- Tenancies of high value dwelling-houses. These are (para 2 & 2A schedule 1 HA 1988):
    - if tenancy or contract for tenancy entered into before 1 April 1990 the rateable value on 31 March 1990 was greater than £1,500 (Greater London) £750 (elsewhere)
    - if tenancy entered into after 1 April 1990 the net rent is greater than £25,000 pa.
- Tenancies at a low rent. These are (para 3 schedule 1, HA 1988):
    - if tenancy or contract for tenancy entered into before 1 April 1990 the net rent is not greater than £1,000 (Greater London) £250 (elsewhere)
    - if tenancy entered into after 1 April 1990 and a rateable value as at 31 March 1990 the net rent is less than two-thirds of the rateable value.

    such low rent tenancies may be protected under Part I LTA 1954.
- Tenancies under Part II LTA 1954 (ie treated as business tenancies) and licensed premises (paras 4 and 5 schedule 1 HA 1988).
- Agricultural land exceeding 2 acres let with dwelling house or an agricultural holding under the Agricultural Holdings Act 1986 (paras 6 and 7 schedule 1 HA 1988).
- Lettings to students (para 8 schedule 1 HA 1988).
- Holiday lets (para 9 schedule 1 HA 1988).
- Tenancies where the landlord occupies as only or principal home at all times since tenancy was granted another part of the same 'building'. Does not apply to purpose built block of flats unless landlord so resident in another part of the same flat (para 10 and Part III schedule 1 HA 1988).
- Property owned by the Crown, local authorities and other statutory bodies (paras 11 and 12 schedule 1 HA 1988).
- Lettings to asylum seekers under Part VI Immigration and Asylum Act 1999.

In addition there are tenancies with enfranchisement or extension rights under the LRA 1967 (as amended) or LRHUDA 1993 (as amended). These are dealt with later in this chapter under the heading Leasehold Enfranchisement.

A further minor group of (formerly controlled) tenancies which are partially business lettings cannot become regulated tenancies (section 24(3) RA 1977) and fall to be considered under Part II LTA 1954.

Each is considered below in terms of the valuation implications of the legislation relating to each category. Those involved with the letting and management of residential property are referred specifically to the above legislation and to the HA Act 1988 and the LTAs of 1985 and 1987.

Assured tenancies

The HA 1980 introduced a new class of tenancy in the private sector known as the assured tenancy. This was intended to encourage the institutional investor to build new homes to let on the open market. Few tenancies were, in fact, created under HA 1980 largely due to the restriction of the provisions to new dwellings provided by 'approved' landlords.

The HA 1988 created a completely new scheme of assured tenancies, borrowing heavily from LTA 1954 and RA 1977. LTA 1954 is paralleled in terms of tenant protection, but grounds for possession are very similar to those found in RA 1977. These assured tenancies may be granted at market rents.

Under HA 1988 all tenancies of residential property created on or after 15 January 1989, other than those statutorily excepted by schedule 1 HA 1988 (see above), were assured or assured shorthold tenancies. Note: The HA 1996 restricts, on a practical level, the creation of assured tenancies, with all tenancies of residential property created on or after 28 February 1997 (other than those statutorily excepted by schedule 1 HA 1988) being assured shorthold tenancies unless a notice is served on the tenant or included in the agreement, that the tenancy is an assured tenancy not an assured shorthold tenancy.

Section 5 HA 1988 specifies that assured tenancies can only be brought to an end with a court order and to obtain such an order the landlord(s) must follow the procedures set out in the Act and specify the ground(s) for possession which must be one or more of those set out in schedule 2 HA 1988, amended HA 1996. In the case of grounds 1–8 inclusive, the court, if satisfied, must order possession; in all other cases the court may order possession.

The valuation of freeholds subject to assured tenancies must be carried out by reference to comparable transactions. The comparables may need to be analysed with reference to either/both capitalisation of income approaches and capital comparison. The valuer needs to be aware that the market may be using a net income approach or, in some places, a more unsophisticated gross income approach. And, if by capital comparison, the market may be working to a percentage of the VP value.

Assured shorthold tenancies (ASTs)

Although HA 1980 introduced the concept of shorthold tenancies, it is HA 1988 and HA 1996 that have developed the concept further.

- From 28 February 1997 all new tenancies (except those in schedule 1 HA 1988) will be ASTs, unless they are specified to be assured tenancies. (Prior to 28 February 1997 a prescribed notice had to be served on the tenant before the grant of the tenancy stating that the tenancy was to be a shorthold tenancy.)
- Prior to 28 February 1997 an AST had to be for a minimum period of six months. From 28 February 1997 an AST can be for less than six months, or on a periodic basis, but the tenant has the right to stay for a minimum period of six months unless one or other of the grounds for possession set out in schedule 2 HA 1988 is established and possession ordered by the court.
- If the AST commenced prior to 28 February 1997 and either the fixed term has ended or the property is on a periodic tenancy, possession (without giving any grounds) can be regained on two months written notice. If the AST commenced on or after 28 February 1997, possession (without giving any grounds) may be regained on two months written notice after either the fixed term comes to an end or at any time during a periodic tenancy, provided it is at least six months from the commencement of the original tenancy.
- There are other (limited) provisions whereby possession may be obtained during a fixed term.

This added ability to recover possession at the end of a fixed term or during a periodic tenancy suggests that such properties will normally be valued on the basis of capitalised term income plus a reversion to vacant possession (VP) capital value or, since the period before obtaining possession can be relatively short, by direct reference to, or discount from, VP values. Individual cases may, however need consideration as to the likelihood of the tenant giving up possession without difficulty or, if not, the time likely to be taken to obtain VP. Here the valuer may require legal advice as to whether the 'accelerated procedure' would be applicable and the likely delay in the courts.

Tenancies subject to the provisions of the Rent Act 1977

The protected or regulated tenancy is one subject to the RA 1977. Its main features are security of tenure and the provision of a 'fair rent'. Security of tenure was achieved by ensuring that once the contractual term came to an end the tenancy effectively continued on the same terms and conditions, but as a statutory periodic tenancy. The most important feature of the 'fair rent' is that the effect on rental value of the scarcity of residential accommodation within an area must be ignored. Thus a 'fair rent' will, in areas of undersupply, be lower than the rent one would expect if the premises were offered in the open market in the absence of the Rent Acts.

Following from HA1988 no new regulated tenancies can be created on or after 15 January 1989. Such tenancies existing before that date continue as regulated tenancies, and can be 'portable' to alternative accommodation provided by the landlord or, in some circumstances, transferable on death. There are, however a number of tenancies which can not be protected tenancies.

In this section the Appropriate Day [AD] is 23 March 1965 or if the property first appears in the Valuation List at a later date, that later date. The relevant Rateable Value [RV] may need to be apportioned on the AD. The first Rateable Value below applies to Greater London, the second to elsewhere.

A tenancy entered into before 1 April 1990 or (if there is a rateable value on 31 March 1990) on or after 1 April 1990 in pursuance of a contract entered into before that date is not protected by RA 1977 if:

- AD on/after 1 April 1973 and RV on AD greater than £1500/750
- AD on/after 22 March 1973 but before 1 April 1973 *and* RV on AD greater than £600/£300 and RV on 1 April 1973 greater than £1500/750
- AD before 22 March 1973 and RV on AD greater than £400/£200 *and* RV on 22 March 1973 greater than £600/£300 *and* RV on 1 April 1973 greater than £1500/£750
- rent is less than two-thirds RV on AD
- AD before 22 March 1973 and RV on AD greater than £400/£200 and rent less than two-thirds RV on 22 March 1973

A tenancy entered into on or after 1 April 1990 (unless, if there is a rateable value on 31 March 1990, in pursuance of a contract entered into before that date) is also not protected if the rent is greater than £1,000 pa (Greater London)/£250 (elsewhere).

The following are also excluded from protection: dwelling houses let with other land; with a payment for board or services; to students; holiday lets; agricultural holdings; licensed premises; resident landlords; Crown, Local Authorities and certain other statutory bodies; when under part II LTA 1954; when an assured tenancy under section 56 HA 1980. (See schedule 15 RA 1977 Act amended HA 1980.)

The RA1965 introduced the concept of 'fair' rent and this is now consolidated in the RA 1977. Section 70 RA 1977 sets out the rules for determining a 'fair' rent. In determining the fair rent, regard is had to all the circumstances, and in particular to the age, character, locality and state of repair, of the dwelling house and, if any furniture is provided for use under the tenancy, the quantity, quality and condition of the furniture. But the following must be disregarded:

(a)   any disrepair or defect attributable to a failure by the tenant to comply with the terms of the tenancy
(b)   any improvement carried out by the tenant otherwise than in pursuance of the terms of the tenancy (renewal of fixtures will be classed as improvements)
(c)   any improvement to the furniture made by the tenant under the regulated tenancy or as the case may be any deterioration in the condition of the furniture due to any ill-treatment by the tenant.

Part IV RA 1977 (as amended) now deals with registration of rent. Landlords and tenants both have the right to apply to the Rent Officer at any time, notwithstanding the existence of a contractual tenancy, for a rent to be registered. On receipt of an application for registration the Rent Officer inspects the premises, usually calls a meeting between landlord and tenant and subsequently notifies the parties of his intention to register a fair rent. Either party may appeal to the Rent Assessment Committee.

Fair rents are registered exclusive of rates and on the assumption that the tenancy is subject to sections 11–14 LTA 1985 (amended section 116 HA 1988 except for tenancies or contracts for tenancies entered into before 15 January 1989). Under these sections, landlords of residential premises let on/after 24 October 1961 for terms of seven years or less are responsible for structural repairs, exterior repairs and repairs to services, water, gas and electricity; this extends to certain fixtures such as WC, baths and basins. The liability of landlords has been further generally extended by the Defective Premises Act 1972.

Once a rent has been registered it remains the maximum recoverable rent for two years save for permitted variations to cover extra rates borne by the landlord and increases where separate service charges exist. Re-application during the two years is only allowed on the grounds that there has been a change in the condition of the dwelling house or the terms of the tenancy. In all other cases the rent remains fixed for two years or until such later time that a rent is re-registered.

As far as the valuation of properties subject to regulated tenancies is concerned, the important points to note are as follows.

1.   The recoverable rent is restricted to the level of a 'fair rent'.
2.   The rent can only be increased after two years.
3.   There are fairly complex rules relating to succession following the death of the original tenant. These are in schedule 1 RA 1977 as amended schedule 4 HA 1988. In brief:

•   if the death of the original tenant was prior to 15 January 1989, then the residing spouse succeeded to the tenancy and if no residing spouse then a member of the family provided 6 months residence immediately prior to the relevant death. On the death of that spouse/ family member, a second succession on the same rules.
•   If the death of the original tenant is on or after 15 January 1989[1] then surviving spouse (or person treated as spouse) if resident before the relevant death becomes a statutory tenant for as long as he or she occupies as his or her dwelling place. This surviving and resident spouse is known as the First Successor. If there is no spouse, then a family member residing with the

---

[1] But subject to transitional arrangements for deaths of original tenants between 15 January 1989 and 14 July 1990

original tenant for two years prior to death can take over the tenancy as an assured tenancy, but is not known as First Successor. On death of the First Successor, tenancy can pass to person who was member of original tenant's family before his death *and* member of First Successor's family immediately before his death *and* residing with First Successor at time of, and for two years prior to, death of First Successor. This person becomes Second Successor on an assured tenancy.

There is still an active market for this type of investment, particularly where a portfolio of such properties is offered for sale. Those active in this market will be aware that a percentage of vacant possession value, between 25% and 50% in most cases, is used to assess market value, when the income approach might only produce realistic market valuations by the use of very low capitalisation rates.

General tenancies excluded from the assured tenancies, assured shorthold tenancies and Rent Acts

As set out above, there are a number of cases where, the tenancy will be held to be outside the RA 1977 or HA1988 and HA 1996 provisions or, provided the correct notices are served at the commencement of the tenancy, the courts have a mandatory power to grant repossession. Examples of such tenancies are: holiday lettings, lettings by educational establishments to students, lettings by absentee owners and lettings of properties purchased for retirement.

If in those cases it is reasonable to assume that vacant possession can be obtained, the property should be valued by direct capital comparison. However, consideration may need to be given to the amount of time and money that may be required to obtain possession, especially if the matter may need to go through the courts.

Tenancies with high rateable values

The total number of tenanted houses and flats falling within this category represents a very small percentage of the whole, but still comprises an important sector of the market, particularly in central London.

To be unprotected and so outside of the RA1977 regime the rateable values or rent must be higher than those set out above, under the heading 'Tenancies subject to the provisions of the Rent Act 1977'.

To be outside of the assured tenancy or AST regime then the rateable values or rent must be higher than those set out above, under the heading 'Private Sector Tenancies'.

The valuation of such properties requires the capitalisation of the current contracted net income, plus, either the reversion to capital value, or to the capitalisation of the reversionary net income, with the reversion and the capitalisation rate being assessed by reference to market evidence. The capitalisation rate may need to reflect the market's view regarding the possibility of such properties becoming assured tenancies.

As many of these properties will be in blocks of flats, special attention must be given to the calculation of landlord's outgoings and in terms of both freehold and long leasehold valuations in or of blocks of flats to the LRHUDA 1993, which gives long leaseholders the right to buy an extended lease or the collective right to buy, under specific conditions, the freehold.

Resident landlords

The RA 1974 introduced a new special class of tenancy for leases granted on or after 14 August 1974 where the landlord resides in the same building. Such tenancies, whether of furnished or unfurnished premises, now fall under section 12 RA 1977. The following must be noted:

- the landlord must be resident at the time of the grant and
- at all times since (other than for 28 days) and
- that if the landlord changes the new landlord must also reside in the building for the resident landlord exemption to continue.

The important feature of a resident landlord letting is that although the tenant enjoys limited protection from eviction (RA 1977, Protection from Eviction Act 1977 amended section 31 HA 1988) the landlord, or on the landlord's death after 28 November 1980, the landlord's personal representatives (if the property is vested in them within two years), can recover possession.

Thus if it is reasonable to assume that vacant possession can be obtained, the property should be valued by direct capital comparison. However, consideration may need to be given to the amount of time and money that may be required to obtain possession, especially if the matter may need to go through the courts. In this connection the landlord only needs to prove that he is resident and that the tenancy has been properly determined, and the court has no option but to order possession. For tenancies that commenced prior to 15 January 1989 section 20 RA 1977 still applies, and the tenancy may be a 'restricted contract'. Should that be the case, then the court has discretion to delay the possession for three months.

If the tenancy was created after 15 January 1989 it will be excluded from being an assured tenancy by schedule 1 (similar provisions apply to circumstances where board and lodging is provided) and will therefore be valued on a vacant possession basis, although the same caveat applies with regard to time and money that may be required to get possession through the courts.

Leasehold reform

Valuers need to consider the Leasehold Reform aspects in the following cases:

- valuations of freehold interests when the tenant has a right to enfranchise or extend its lease
- valuation of long leases where the tenant has a right to enfranchise or extend the lease
- valuation of price of enfranchisement or lease extension for freeholders or leaseholders.

Strictly speaking the 'lease extension' is a surrender of the existing lease and grant of a new lease, with the new lease being for the previous remaining term plus the period of extension set out by statute. The term 'lease extension' or 'extended lease' is, however, generally used, and will be so used here.

Crown property

Crown property is, strictly speaking, exempt from LRA 1967, LHRUDA 1993 and CLRA 2002. However, by undertakings given to both Houses of Parliament (Sir George Young to the Commons, Hansard 2 November 1992 and Lord Falconer to the Lords Hansard 19 November 2001), the Crown has agreed to act by analogy to the statutes with certain limited exceptions. Where the exceptions apply such that the freehold will not be sold, the Crown will be prepared to negotiate new leases. Since the Crown is acting by analogy with the statute, the following concepts and valuation methods should apply as if it were acting under statute.

Houses

Almost all long lessees of houses satisfying a number of qualifying conditions now enjoy the right to acquire the freehold title to their home — the right to enfranchise. Some have a right to request a 50 year extension to their lease.

This may include a property that may reasonably be called something else (eg shop with flat over) as well as reasonably being called a house (eg house, part of which is used for retail purposes). But note also the restriction (below) on leases to which Part II LTA 1954 applies.

All tenants seeking to enfranchise must satisfy the following conditions:

- Must not be an excluded tenancy under section 1(3) LRA 1967 (house ancillary to letting of land; some agricultural tenancies; some tenancies from charitable housing trusts)
- Must be a 'long tenancy'— ie
  - (i)   originally granted for term certain of greater than 21 years subject to certain exceptions for leases terminable after death or marriage [section 141 and schedule 2 HA1980]
  - (ii) tenures that although less than 21 years are considered to be long tenures [under section 149(6) LPA 1925; perpetual renewal by section 3(1) LRA1967; non perpetual renewal without premium by section 3(4) LRA1967; some 'right to buy' under Part V of HA1985; 'shared ownership' lease under section 143 HA1985; some continuation tenancies under Part I or II LTA 1954]
  - (iii) tenancies to which Part II LTA 1954 apply are excluded unless they are for a term certain exceeding 35 years or are within the first three types mentioned in para (ii) immediately above.

Valuers need to be particularly aware of subsections 3(2) LRA1967, enabling tenancies to be joined together. This is of especial importance in considering the issue of 'tenant's improvements' carried out under the tenancy.

- The tenant must be the tenant of the whole (or tenant of part and already freeholder of remainder) for the last two years for claims on or after 26 July 2002.
- Must be at a 'low rent' [section 1(1) LRA1967].

    There are three ways in which the 'low rent' test may be satisfied. The legislation contains various exceptions and possible ambiguities and valuers are strongly advised to obtain the client's solicitors' advice before preparing valuations based on expectation of enfranchisement.

    Valuers also need to be aware that both the claim for the extended lease under section 14 LRA 1967 and the 'site value' valuation under section 9(1) LRA 1967 only apply where the 'low rent' test under section 4(1) LRA 1967 is satisfied.

The 1967 Act 'low rent' test:
- for tenancies commencing after 31 August 1939 and before 1 April 1963 rent not to exceed two thirds of the letting value of the property
- for other tenancies/contract for tenancies commenced prior to 1 April 1990 the property is at a 'low rent' at any time when rent does not exceed two thirds of the rateable value of the property on the 'appropriate day' or, if later, first day of the term. The 'appropriate day' is normally 23 March 1965 unless (generally speaking) it first appears in the Valuation List after that date

– for tenancies commenced on or after 1 April 1990, rent not to exceed £1,000 (Greater London) or £250 (elsewhere).

If the 1967 Act test is failed, then the 1993 Act test is used:

– for tenancies granted prior to 1 April 1963 the rent must not exceed two thirds of the letting value of the property
– for tenancies granted on/after 1 April 1963 but before 1 April 1990 (or by contract for tenancy before 1 April 1990), rent must not exceed two thirds of rateable value on the 'relevant date'. 'Relevant date' is date of commencement of tenancy or, if no or nil rateable value on that day, the date on which it first had a rateable value other than nil
– for tenancies granted on or after 1 April 1990 (except due to a contract before that date or where no rateable value prior to 1 April 1990) rent not to exceed £1,000 (Greater London) £250 (elsewhere).

If the 1993 Act test is failed, then under section 1AA LRA 1967 it is deemed to be at a low rent unless it is an 'excluded tenancy'. Excluded tenancies are those in a rural area designated for this purpose, where the freehold is owned with adjoining land since 1 April 1997 and the adjoining land is not occupied for residential purposes and the tenancy was either granted before 1 April 1997; or for a term not exceeding 35 years between 1 April 1997 and 26 July 2002 (England) 1 Jan 2003 (Wales).

Enfranchisement price

There are three different methods of calculating the enfranchisement price under the 1967 Act, which is used depends upon the basis of qualification of the property for enfranchisement.

(1) Section 9(1) — also known as 'the original basis' or 'the 1967 Act basis'
(2) Section 9(1A) — also known as 'the 1974 Act basis' as introduced by HA1974
(3) Section 9(1C) — introduced by LRHUDA 1993.

(1) Section 9(1) — as amended S82 HA1969 to remove 'the tenant's bid' (ie no marriage value)

Qualification conditions: in order to be valued under section 9(1):

(i) Not a lease granted under 'right to buy' of part V HA 1985. This is valued under section 9(1A) LRA1967
(ii) Not a 50-year extension lease granted under section 14 LRA1967 This is valued under section 9(1C) LRA1967
(iii) Rateable value
   (a) if the tenancy/contract for tenancy entered into prior to 1 April 1990 *and* the appropriate day before 1 April 1973 the property must have rateable value on appropriate day not exceeding £400 (Greater London) £200 (elsewhere)
   (b) if tenancy/contract entered into before 1 April 1990 and house had rateable value at commencement of tenancy or otherwise before 1 April 1990, rateable value on appropriate day not to exceed £400 (Greater London) £200 (elsewhere) or
   (c) if tenancy created on or before 18 February 1966 but appropriate day on or after 1 April 1973, then rateable value on appropriate day not to exceed £1,500 (Greater London) £750 (elsewhere) or

(d) if tenancy created after 18 February 1966 but appropriate day on or after 1 April 1973, the rateable value on appropriate day not to exceed £1,000 (Greater London) £500 (elsewhere)

(e) if tenancy created on or before 18 February 1966, and rateable value on appropriate day greater than £400 (Greater London), £200 (elsewhere), the rateable value on 1 April 1973 not to exceed £1,500 (Greater London), £750 (elsewhere)

(f) in addition to satisfying one or other of the rateable value tests above, the rateable value cannot be above £1,000 (Greater London), £500 (elsewhere)on 31 March 1990 after adjusting rateable value to take account of tenant's improvements (if it is, section 9(1A) will apply)

(g) if there is no rateable value on 31 March 1990 the R under the formula in section 1(1)(a)(ii) of 1967 Act must not exceed £16,333.

Calculation of price

The price to be paid on enfranchisement under section 9(1) LRA1967 is the market value of the landlord's interest on the assumption that the tenant had acquired a 50 year extension of his lease under section 14 LRA1967. Section 82 of the Housing Act 1969 amended section 9(1) so that the tenant's own bid for the freehold is to be disregarded.

Thus, those aspects which are required to be found are the term rent, the Modern Ground Rent [MGR] required by section 14, the value of the final reversion and the relevant yield for the transaction. The MGR is normally taken as a percentage of the site value, thus requiring ascertaining site value. Although this can be done by comparable sites, it is normally (especially in urban areas) by taking a percentage of the value of the property that is already there or that might be built there if the site was vacant. The valuer should note that the section 14 lease is on the 'same terms as the existing tenancy' and the MGR is for the 'uses to which the house has been put' other than uses that are 'not permitted' or 'permitted only with landlord's consent' [section 15 LRA1967]. Where the existing lease has a long unexpired term, then the final reversion (ie plus 50 years) is normally taken by capitalising the MGR in perpetuity.

Example 10.1 Section 9(1) valuation with substantial term remaining

House on lease with 58 years remaining at fixed rent of £10 pa

Equivalent freehold properties in modernised condition selling at £500,000

Ground rent		£10	
YP 58 yrs at 7%		£14.00	140
Reversion to	£500,000		
Site Value at	35%		
Site Value	£175,000		
Ground Rent at	7%		
Modern Ground Rent		12,250	
PV in perp def 58 yrs at 7%		0.28226	3,458
			£3598 say £3,600

In view of the use of the same yield for decapitalisation and capitalisation of the Modern Ground Rent this can be simplified to

Ground Rent	£10	
YP 58 yrs at 7%	14.00	140
Reversion to	£500,000	
Site Value at	35%	
	175,000	
PV 58 yrs at 7%	0.1976	3,458
	£3,598 say £3,600	

Example 10.2 Section 9(1) valuation with short term remaining

House on lease with 8 years remaining at fixed rent of £10 pa.
Equivalent properties in modernised condition selling at £500,000; this property in good repair, but unmodernised is worth £400,000 at valuation date.

Ground Rent	£10	
YP 8 yrs @ 7%	5.97	60
Value of 'modern equivalent'	£500,000	
Site Value at	35%	
	£175,000	
Ground Rent at	7%	
	£12,250[1]	
YP 50 yrs at 7%	13.80	
PV 8 yrs at 7%	0.582	98,394
Final Reversion	£400,000[2]	
YP 58 yrs at 7%	0.01976	
		7,903
	£106,357 say £106,360	

Notes
(1)   Valuers may wish to consider as to whether such a 50 year site lease would ever be let at the figures thrown up by this form of valuation .
(2)   There is also an argument that the property may, or may not, be 'modernised' or 'improved' within this 58 year time span and if so, whether the final reversion figures or yield rate here should be amended for this aspect.

(2) Section 9(1A) — 'the 1974 Act basis'

Following the introduction of the 1973 Valuation List, the Housing Act 1974 brought more properties into the enfranchisement procedure. In essence, if a house is enfranchiseable, but not valued under section 9(1) — see above — it is valued under section 9(1A) unless it is one of those to which section 9(1C) applies.
    Assumptions to note:

(i)   tenant has right to remain in possession of the property at end of tenancy (now generally under schedule 10 HA 1989)

(ii) tenant has no liability to carry out repairs, redecorations under the lease or Part I LTA 1954
(iii) price diminished by increase in value due to improvements carried out by tenant or predecessor in title at their own expense
(iv) various rent charges and rights, as set out in 9(1A)(e) and (f) as appropriate.

Points for the valuer to note

(a) Some properties only appear to fall within section 9(1A) due to the rateable value on 31 March 1990 being greater than £1,000 (greater London) £500 (elsewhere). Consideration must be given as to whether a notional reduction in rateable value is due to tenant's improvements.
(b) In addition, under subsections 3(2) and 3(3) LRA 1967, leases may be joined together to form one long lease. Improvements carried out by the tenant under the earlier leases should be taken into account, with regard to considering the correct rateable value and/or in the calculation of price. This may involve detailed research into the history of the building, using local studies libraries, building control, former tenants or their architects etc.
(c) Removal of the value of the tenant's improvements may also require adjustment of the current or reviewed ground rent, where the ground rent is expressed to be a percentage of the 'rack rent' or related to the freehold value of the property.
(d) Frequently the existing lease value is expressed as a percentage of the freehold value.

Example 10.3 Section 9(1A) valuation

House on lease with 58 years remaining, at fixed rent of £10 pa for 8 years, then review to 1% of freehold value with vacant possession. Value of property as a freehold in its current condition is £500,000, and as a leasehold £375,000 (75%).

Term				
Ground Rent		£10		
YP 8 yrs @ 7%		5.97	60	
Review to	Freehold Value	500,000		
	Rent at	1%		
		5,000		
	YP 50 yrs @ 7%	13.80		
	PV 8 yrs @ 7%	0.582	40,158	
Reversion to		500,000		
	YP 58 yrs @ 7%	0.01976	9,880	50,098
Marriage Value				
After	Landlord	nil		
	Tenant	500,000		
Total		500,000		
Before	Landlord	50,098		
	Tenant	375,000	425,098	
Marriage Value			74,902	
Half share to Landlord at			0.5	
				37,451
				£87,549 say £87,550

Example 10.4 Section 9(1A) valuation

Same facts as before but:

(i)   on investigation it is found that there was a prior lease dating back to the late 1800s, which was surrendered and renewed by the then tenant for the current lease

(ii) on further investigation it is found that significant improvements were carried out under both the first lease (eg installation of all internal bathrooms) and second lease (eg extra storey)

(iii) the value attributable to each set of improvements is £50,000, making £100,000 in all on a freehold basis.

Term					
Current Ground Rent			£10		
YP 8 yrs @ 7%			5.97	60	
Review to	Freehold Value		£500,000		
	Less improvements		100,000		
			400,000		
	Ground rent at		1%		
			4,000		
	YP 50 yrs @ 7%		13.80		
	PV 8 yrs @ 7%		0.582	32,126	
Reversion to			£500,000		
	Less improvements		£100,000		
			£400,000		
	PV 58 yrs @ 7%		0.01976	7,904	
Landlord's current interest					40,090
Marriage Value					
After	Landlord		nil		
	Tenant		500,000		
	Less improvements		100,000	400,000	
Before	Landlord		40,090		
	Tenant	375,000			
Less improvements		75,000	300,000	340,090	
Marriage value				59,910	
Half share to Landlord				0.5	29,955
					£70,045

(3) Section 9(1C) — the 1993 Act basis

The LRHUDA 1993 and the CLRA 2002 further extended enfranchisement to houses, in particular bringing the following in:

- properties that failed to meet the 1967 Act low rent test but met the 1993 Act low rent test
- properties that failed the 1993 Act low rent test but were not excluded tenancies
- certain tenancies and contracts for leases entered into before 18 April 1990 but terminable by death or marriage

- leases extended under section 14 LRA 1967 where term date of original lease has passed.

To these properties the section 9(1C) valuation basis applies.

This is the same as the section 9(1A) basis with an added amount (if any) for any compensation due to the landlord in relation to:

(i)   any diminution in value of other property due to this acquisition
(ii) any other loss or damage (including loss of development value in relation to the house and premises) due to the acquisition to the extent it is referable to his ownership of any other interest in other property.

Flats: Extension of leases and collective enfranchisement

LRHUDA 1993 and CLRA 2002 extended the enfranchisement provisions to leases in two distinct ways.

(a)   enabling individual leaseholders of flats to extend their leases for 90 years, with the new lease being granted on, effectively, the same terms and conditions as before, except at a peppercorn ground rent
(b)   enabling a participating group of leaseholders of flats in a 'block' to purchase the freehold of that property, providing certain prior conditions are met.

Extension of leases:

The prior requirements are:

(a)   a lease which is for term of years certainly exceeding 21 years (or has become such)
(b)   not a business lease
(c)   not granted by charitable housing trust in pursuit of its charitable objectives
(d)   not granted in breach of terms of superior lease and breach not waived
(e)   lease owned for at least two years prior to a claim for extended lease being made.

Process

The full process goes beyond the purview of this book, and is for the client's solicitors to follow through. In brief:

(a)   the Notice of Claim by tenant is prepared and served specifying date for counter notice
(b)   the counter notice by landlord is prepared and served
(c)   time for negotiations is allowed
(d)   application to Leasehold Valuation Tribunal (LVT) for determination of price and other terms may be made between two and six months after the date the counter notice was given
(e)   if negotiations are not concluded between the parties the matter will be listed and heard by the LVT with right of appeal to the Lands Tribunal.

Points for the valuer to note about the process:

(a) Although the figure inserted in the Notice of Claim may be, effectively, a low opening offer, it must still be a realistic figure. Failure for it to be a realistic figure will invalidate the notice. This may have serious consequences in at least three particular cases: leases with just over 80 years to expire; leases that are in the process of being sold; leases in rapidly rising markets.

(b) The time-limits for service of Counter Notice and applications to LVT or court are absolutely strict. When receiving instructions to prepare valuation for the Notice it is important to check if the 80 year mark is close; and for the Counter Notice it is imperative to ensure the date by which it must be provided.

(c) Failure to serve a Counter Notice (or valid Counter Notice) will generally lead to an application by the tenant to the county court for an order determining the acquisition of the new lease on the basis set out in the Notice of Claim. Provided the Notice of Claim is valid and properly served the court has no discretion but must grant such an order.

d) If there is a head leasehold interest, or other intermediate leases, as well as a freehold interest, each of these may need to be valued separately. The valuer should seek his client's solicitors' instructions.

Valuation

In essence the valuation consists of three parts, although in most cases only two of these are used and these are, themselves, run together into one valuation. They are (all refs to LRHUDA1993):

Schedule 13 para 3	diminution of value of landlord's interest in the flat. This is, effectively, the loss of the ground rent for the remaining term of the existing lease, and the loss of the reversion thereafter. Normally it is calculated by a fairly standard 'term and reversion' method, as the landlord's interest after the transaction is usually nil. The landlord's value after the transaction may need to be re-considered, especially if the existing lease is very short and the value of the extended lease is very high.
Schedule 13 para 4	half 'marriage value'. Marriage value is calculated as difference between the total of the interests after the transaction is done and the total of those interests before. Where the existing lease has more than 80 years to run, the marriage value is taken to be nil.
Schedule 13 para 5	compensation in relation to diminution in value of the other properties and other loss or damage (including loss of development value in tenant's flats).

Points for the valuer to note about the valuation:

(a) Valuers need to take particular care with valuations where the lease is approaching 80 years, and especially when within two months of that cut off point. For if a figure that is held to be 'unrealistic' is supplied for the Notice of Claim, the landlord may choose to wait until the 80 year point is passed before raising the issue of the validity of the Notice, leaving a new, higher figure having to be paid due to the inclusion of half marriage value.

(b) Similarly, if an 'unrealistic' valuation is supplied for a Notice of Claim on a lease which is to be assigned (together with that Notice), then the issue of validity of notice may not be raised until after the assignment has taken place — leaving the new owner to wait two years before purchasing the extended lease at (probably) an increased figure.

(c) At the time of writing there is a considerable debate about the relevant yield(s) to be used in the 'term and reversion' valuation under schedule 13 para 3, and the method by which this/these should be determined or calculated.

(d) All valuations should be carried out in a 'no Act world' — that is, as if the LRHUDA 1993 and CLRA 2002 had not been passed and were not in existence.

(e) Any increase in value due to tenant's improvements is to be disregarded.

(f) There are cases where the rent currently paid by the individual flats to a headlessee equals the rent paid by the headlessee to the freeholder. (This is normally in the case of a management company as headlessee.) The replacement on one flat of the current rent by a peppercorn rent puts the headlessee in a deficit position, and particular care must be taken in the valuation of such a negative interest (see *Blackmun* v *Trustees of Portman Re Wendover Court 2003* — LEASE LVT Decisions 1993 Case 566). Alternates, (but by agreement only) are that the flat continues to pay the rent(s) previously reserved for the length of the existing term but no or reduced compensation to the headlessee; or that the freeholder reduces the rent receivable under the headlease and receives additional compensation in respect of that.

(g) Substantial information about previous LVT cases (including downloads of the cases) is available from LEASE at *www.lease-advice.org*. Care needs to be taken when using the table of LVT Decisions 1993 Act on that site, as the brief summary does not always reflect the actual decision, and must be checked before production to any court or tribunal.

Example 10.5 Schedule 13 valuation — individual flat

Flat on lease with 58 years remaining, at fixed rent of £100 pa for 9 years then review to £200 for 25 years then review to £300 for final 24 years. Value of property as a long leasehold in its unimproved condition, but in repair, £150,000, and as a leasehold £132,000 (88%).

Term			
Ground Rent		£100	
	YP 9 yrs at 7%	6.51	651
Review to Ground Rent		£200	
	YP 25 yrs @ 7%	11.65	
	PV 9 years @7%	0.544	1,268
Review to Ground Rent		£300	
	YP 24 yrs at 7%	11.47	
	PV 54 years at 7%	0.026	89
Reversion to		150,000	
	YP 58 yrs at 7%	0.01976	2,964
Landlord's Current Interest			4,972
Marriage Value			
After	Landlord	nil	
	Tenant	150,000	
Total			150,000

Before	Landlord		4,972	
	Tenant		132,000	136,972
	Marriage Value			13,028
Half share to Landlord at			0.5	6,514
				11,486
				say £11,485

Collective enfranchisement

The prior requirements are:

(a) the building is a structurally detached self contained building or part of a building that is a vertical division of the building that can be redeveloped independently of the remainder of the building and where the services are provided independently or could be provided without significant disruption to others in the building

(b) those parts not occupied for residential purposes nor comprised in the common parts of the building, do not exceed 25% of the internal floor area of the whole

(c) at least two thirds of the flats in the 'block' must be 'qualifying' leases ie for more than 21 years which are not business leases, granted by a charitable housing trust in pursuit of its charitable objectives, or granted in breach of superior lease with the breach not being waived

(d) the number of qualifying tenants participating must be at least half of the total number of flats in the block, except where two flats only when it must be both of them

(e) the block is neither excluded under the 'resident landlord' exemption — see section 10 L&HUDA 1993 as amended section 118 CLRA 2002, nor as part of a railway.

Process and points for the valuer to note about the process.

In general the process and points to note are as for Lease Extensions (see above).

There are, however, three additional issues:

(i) In an estate where more than one block wishes to enfranchise, each block must be dealt with and valued, separately. This is even if the participating group of each block wish to keep the Estate together by enfranchising together. Each group may, however, nominate the same person as 'Nominee Purchaser' eg XXX Estate (Freehold) Ltd.

(ii) In the Notice of Claim for Freehold, both the 'specified premises' (meaning the actual building and attached structure housing the flats in question), and the common and appurtenant land are required by LRHUDA 1993 to be specified, both in the notice and on the attached plan, and also valued separately. There is judicial precedent for suggesting that this is no longer necessary but the valuer should be aware of the issue and take appropriate instructions.

(iii) Similarly, if there is a head leasehold interest, or other intermediate leases, as well as a freehold interest, each of these must be specified, and valued, separately.

Thus in a relatively simple case of a block of flats with surrounding gardens, where there is a freeholder and headleaseholder (eg management company), there are *four* interests to be valued, all of which are by reference to schedule 6 (see below).

Valuation

Although schedule 6 LRHUDA 1993 follows the same basic three part format for collective enfranchisement valuations as does schedule 13 for individual lease extensions, the schedule 6 valuations can be considerably more complex with more traps for the unwary.

For the valuation of the freehold the three basic elements are:

Schedule 6 para 3   value of freeholder's interest excluding tenants' bid on following assumptions:
- fee simple subject to any leases that are to be acquired
- 'no Act world' — but see below under 'hope value'
- disregard participating tenants' improvements
- include other rights and burdens specified

Schedule 6 para 4   unless more than 80 years to run, calculation of half marriage value, with marriage value being effectively the aggregate of the property interests after the transaction less the aggregate of the property interests before the transaction, on the same assumptions set out immediately above

Schedule 6 para 5   compensation in relation to diminution in value of the other properties and other loss or damage (including loss of development value in participants' flats).

Points for the valuer to note:

(a) In the para 3 valuation, tenants' improvements carried out by non-participating tenants are not disregarded.
(b) In the para 4 valuation, there is no 'marriage value' attributable to non-participating flats. An argument has been advanced that the valuation should reflect 'hope value' in that the participating group may 'hope' that the non-participants will extend their leases at a future date. This will depend upon the facts of the individual case.
(c) There is no minimum ownership period for participation in collective enfranchisement.
(d) Particular attention must be paid to a valuation for Notice of Claim where any participants are close to (and especially within the last two months or so) of the 80 year period, to ensure the valuation in the Notice cannot be considered as unrealistic.
(e) In blocks with non-participants, the present value of the freehold due to non-participants' flats is sometimes erroneously omitted from the 'after' part of the marriage value calculation.
(f) Thus, especially with larger blocks, where there may be a variety of lease terms, rents and review dates, the valuer should consider doing a check valuation on a flat by flat basis, using para 3 only for non-participants and paras 3 and 4 (as for lease extension) for participants. Due to the mathematics, this is equal to the sum of the formal para 3 and para 4 valuation required by statute.
(g) At the time of writing, there is a considerable debate about the yield(s) to be adopted for term and reversion, and the methods by which they are calculated or derived.
(h) Substantial information about previous LVT cases (including downloads of the cases) is available from LEASE at *www.lease-advice.org*. Care needs to be taken when using the LVT Decisions 1993 Act Table on that site, as the brief summary does not always reflect the actual decision, and must be checked before production to any Court or Tribunal.

Example 10.6 Schedule 6 valuation — freeholder only, with no headlease

Block of 36 flats, with two types of flats, and no common or appurtenant land. 20 Flats are of Type A, held on leases with 58 years remaining at a fixed ground rent of £10 pa. Type B have 79 years remaining at a current rent of £50 pa, rising by £50 in 5 years and every 25 years thereafter. The current improved value of Type A as a 'freehold equivalent' is £150,000, and as a 58 year lease, £132,000. The value of Type B as a 'freehold equivalent' is £180,000, and as a 79 year lease £177,000. All Type A flats have been improved by their tenants by installation of central heating systems, which have added £5,000 to the value of both freehold equivalent and existing lease values. 30 flats (18 of Type A and 12 of Type B) are participating in the collective enfranchisement.

Formal Schedule 6 Valuation
Landlords current interest:

Type A Term

Ground Rent	20 flats at £10	£200		
YP 58 yrs at 7%		14.00		2,800
Reversion to Participants 'A'				
Flat value		150,000		
Less Improvements		5,000		
Unimproved Value		145,000		
No of Participants		18	2,610,000	
Reversion to Non-Participants 'A'				
Flat value		150,000		
No of Non-Participants		2	300,000	
Total Type A			2,910,000	
PV 58 yrs at 7%			0.01976	
				57,502
Type B				
Term				
Ground Rent	16 flats at £50	£800		
YP 5 yrs at 7%		6.51	5,208	
Review to Ground Rent		£1,600		
YP 25 yrs at 7%		11.65		
PV 5 yrs at 7%		0.713	13,290	
Review to Ground Rent		£2,400		
YP 25 yrs at 7%		11.65		
PV 30 yrs at 7%		0.131	3,663	
Review to Ground Rent		£3,200		
YP 24 yrs at 7%		11.47		
PV 55 yrs at 7%		0.024	881	
Total Term 'B'				23,042
Reversion to All 'B'				
Flat value		180,000		
No of Flats		16		
Total Type B			2,880,000	
PV 79 yrs at 7%			0.004772	13,743
Landlord's Current Interest				97,087

Marriage Value

After	Landlord			nil	
	Participants 'A'				
		Flat value	150,000		
		Less Improvements	5,000		
		Unimproved Value	145,000		
		No of Participants	18	2,610,000	
	Participants 'B'				
		Flat value	180,000		
		No of Flats	12		
				2,160,000	
	Non-Participants A				
		Leases 2 at £132,000		264,000	
	Non Participants B				
		Leases 4 at £177,000		708,000	
	Freehold Value of Non Participants				
		2 of Flat A			
		Term			
		Ground Rent 2 flats at £10	£20		
		YP 58 yrs at 7%	14.00	280	
		Reversion			
		Flat value	150,000		
		No of Non-Participants	2		
			300,000		
		PV 58 yrs at 7%	0.01976		5,928
		4 of Type B			
		Term			
		Ground Rent 4 flats at £50	£200		
		YP 5 yrs at 7%	6.51	1,302	
		Review to Ground Rent	£400		
		YP 25 yrs at 7%	11.65		
		PV 5 yrs at 7%	0.713	3,323	
		Review to Ground Rent	£600		
		YP 25 yrs at 7%	11.65		
		PV 30 yrs at 7%	0.131	916	
		Review to Ground Rent	£800		
		YP 24 yrs at 7%	11.47		
		PV 55 yrs at 7%	0.024	220	
		Reversion			
		Flat value	180,000		
		No of Non Participants	4		
		Total Type B	720,000		
		PV 79 yrs at 7%	0.004772	3,436	

Total After Transaction   5,757,405

Before	Landlord			97,087
	Participants A			
	Lease Value	132,000		
	Less Improvements	5,000		
		127,000		
	No of Participants	18	2,286,000	
	Non-Participants Flat A			
	Lease Value	132,000		
	No of Non Participants	2	264,000	
	All B			
	Lease Value	177,000		
	No of Flats	16	2,832,000	
Total Before Transaction			5,479,087	

	278,318
Half share to Landlord at	× 0.5
	139,159
Landlord's Current Interest	97,087
	236,246.

Check Valuation (and shortcut method)

Flat A — as Non Participant

Term: Ground Rent £10 × YP 58 yrs at 7% (14.00)	140
Reversion to £150,000 × PV 58 yrs at 7% (0.01976)	2,964
Landlord's Current Interest	3,104
No of Non Participants 'A' Payable to Freeholder	2

6,208

Flat A — as Participant

Term: Ground Rent £10 × YP 58 yrs at 7% (14.00)	140
Reversion to unimproved £145,000 × PV 58 yrs at 7% (0.01976)	2,865
Landlord's current interest	3,005

Marriage Value

Unimproved Value after Transaction	145,000	
Landlord's Current Interest	3,005	
Tenant's Current Unimproved Interest	127,000	
Total Current Interests	130,005	
Marriage Value	14,995	
Half Share to Landlord	0.5	7,497.5
		10,502.5
No of Participants 'A'		18
Payable to Freeholder		189,045

Flat B — as Non Participant

Term

Ground Rent	£50	
YP 5 yrs at 7%	6.51	326

Review to Ground Rent	£100	
YP 25 yrs at 7%	11.65	
PV 5 yrs at 7%	0.713	831

Review to Ground Rent	£150	
YP 25 yrs at 7%	11.65	
PV 30 yrs at 7%	0.131	229

Review to Ground Rent	£200	
YP 24 yrs at 7%	11.47	
PV 55 yrs at 7%	0.024	55

Reversion

Flat value	180,000	
PV 79 yrs at 7%	0.004772	

		859
Landlord's Current Interest		2,300
No of Non Participants		4
Payable to Freeholder		9,200

Flat B — as Participant

Landlord's Current Interest per flat (from immediately above)		2,300

Marriage Value

Unimproved Value after Transaction		180,000	
Landlord's Current Interest	2,300		
Tenant's Current Interest	177,000		
Total Current Interests		179,300	
Marriage Value		700	
Half Share to Landlord		0.5	350

		2,650
No of Participants 'B'		12
Payable to Freeholder		31,800
Total payable to Freeholder		236,253

The difference between this and the £236,246 of the formal valuation is due to minor rounding errors. In both methods the figure would probably be rounded to £236,250 although the statute does not allow for rounding.

Valuation of intermediate leasehold interests

1 *If a 'minor intermediate lease'* being a lease with an expectation of reversion of less than one month and a profit rent of not more than £5 pa then

$$\text{Price paid} \quad = \quad \frac{R}{Y} \quad - \quad \frac{R}{Y(1+Y)^n}$$

Where:

R = profit rent

Y = yield on 2.5% Consolidated Stock

n = length remaining (rounded up to whole year) of minor intermediate lease

2 *If not a minor intermediate lease,* valuation follows the format of the para 3 valuation (set out above) subject to such modifications as are appropriate.

Where marriage value is payable to the freeholder, that marriage value is split between the freeholder and intermediate leaseholder(s) in proportion to the value of their respective interests (as calculated under, effectively, para 3 of schedule 6 or 13).

Example 10.7 Brief example of intermediate lease calculation

Assume calculation under para 3 of schedule 6 or schedule 13 of diminution of value of freeholder's interest is £1,000.
    Para 3 of schedule 6 (or schedule 13 for lease extensions) is done using the normal term and reversion method, and was set out in full in the preceding sections above, both with regard to 'extension of leases' and 'collective enfranchisement'.

Assume calculation under para 3 of diminution of value of headlessee's interest is £100.
Again, para 3 of schedule 6 (or schedule 13 for lease extensions) is done using the normal term and reversion method, and was set out in full in the preceding sections above, both with regard to to 'extension of leases' and 'collective enfranchisement'. Frequently with Headleases (although not always) there is no reversionary value.

Assume value of flat is £150,000 on freehold equivalent basis, and £132,000 on current lease:

Valuation

Freeholder's Para 3 figure			1.000
Headlessee's Para 3 figure			100
Total Landlords' interests			1,100
Marriage Value			
Total Value after			150,000
Value before	Freeholder	1,000	
	Head lessee	100	
	Leaseholder	132,000	
Total current interest			133,100
Marriage Value			16,900
Half Marriage Value to Landlords			0.5 8,450
Total payable by Leaseholder			9,550

Of which Freeholder receives $\dfrac{1,000}{1,100} \times 8,450$ plus 1,000   = 8,682

Of which Head lessee receives $\dfrac{100}{1,100} \times 8,450$ plus 100   = 868

Total   9,550

Development Opportunities 11

In relation to the 'income approach' residual valuations and development appraisals require the valuer to assess and capitalise the potential income stream from a proposed development which does not yet exist. In many cases development is 'forward funded' using long term finance from an investor such as a Pension Fund or a consortium of private and/or corporate investors.

This may occur:

- in the case of a bare or underdeveloped site where planning permission for development has been, or is likely to be, obtained
- in the case of an existing building for refurbishment or where planning permission has been, or is likely to be, obtained for a refurbishment and/or change of use
- in the case of an existing building where planning permission has been, or is likely to be, granted for its demolition and replacement.

This capital facilitates:

- acquisition of the site
- the provision of the infrastructure
- the completion of construction
- all professional fees in connection with securing of detailed planning and building regulation permission, finance, legal, marketing, design, engineering, land contamination and remediation (if any) etc
- cost of short term financing and
- sales and marketing of the finished product.

The income approach is therefore required to capitalise the potential future income as part of a risk and return analysis undertaken often at the site acquisition stage, while the direct comparison method is preferred and should be used as a 'reality check' when assessing site values. However, it is in the market for property with development potential that the most difficulty in finding good comparable evidence is experienced. In the investment market, the income approach is the most suitable and widely used alternative to direct capital comparison.

Important Health Warning:

- *Development Appraisals* (where the profit from a scheme is assessed using an actual or estimated land price) and
- *Residual Valuations* (where the land value is assessed using a pre-determined amount of developers profit).

are notoriously unreliable and have been criticised by the Lands Tribunal and in Court Judgements. The adage 'garbage in-garbage out' is particularly pertinent to these appraisals as:

(a) *they involve an element of forecasting*, as the calculations are undertaken sometime before the development is completed and lettings or sales are completed. In the case of complicated development projects this may be some years into the future due to the need to obtain planning consents, complete infrastructure works and undertake building operations.
(b) *they are complex* and involve many variables which are highly sensitive to changes in both macro-economic conditions and local micro-economic factors.

A full consideration of the development process is outside the scope of this book, however before considering the income approach in relation to development opportunities we feel it is wise to illustrate a preferred methodology to development appraisal. This process is a *sifting* methodology which filters potential uses and is a pre-cursor to the financial analysis of a development scheme for a particular use or combination of uses.

The highest and best use methodology

A development appraisal can be required in a number of different circumstances, and depending on these, the emphasis given to various aspects of the solution will vary. The most common development scenarios are where a developer owns or has an option on a site (which may be greenfield, brownfield, cleared or contain existing buildings) and we wish to determine the optimum use for it.

Conversely we may have a development scheme in mind for which we know there is a proven market demand (eg a development of up-market apartments or a mixed use scheme) but the developer needs to find the best location. Although differing in certain aspects, these are two sides of the same coin and both involve a good deal of similar work. Crucial to both sets of circumstances is the recognition of the need to find the optimum use for specific sites or locations. In North American appraisal practice this concept of 'optimum' or 'highest and best use' is the litmus test against which all proposals are judged. Essentially North American Appraisal Regulations require the valuer or appraiser to advise the client of the most profitable concept which is feasible for a given site. That concept of feasibility is judged against three criteria:

- technical feasibility
- legal feasibility
- market feasibility.

In the UK we are not quite so explicit in our view, but the old adage that the success of a development rests on location, location and location, does express the reverse of the coin. The task is to tie these concepts together in a rational and convincing manner.

The method

The highest and best use represents an explicit, logical and robust approach to refining the potential uses for a site.

In theory any given site has a range of possible uses. The extent of this range is limited by various considerations such as site area, accessibility, slope, stability and surrounding uses. It may be quite possible therefore to build housing, shops, leisure facilities commercial or industrial, in fact any number or mix of uses on the site.

It should be recognised that:

*   some of the proposed uses will produce a higher rate of return than others
*   not all represent feasible solutions in this location
*   not all types of development will be granted planning permission.

Appraisers need therefore to refine their list of possible solutions to take these conditions into account. Figure 11.1 below demonstrates the process. The outer bold rectangle illustrates all the possible uses of the site.

Figure 11.1 How we refine the highest and best choices from the list of all possible uses

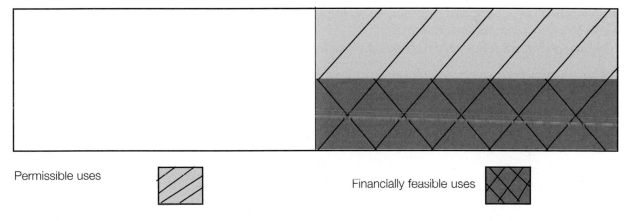

Permissible uses

Financially feasible uses

Starting from the total possible uses (the largest box) we refine our thinking first by identifying those possible uses which are permissible (the two right hand boxes). This process involves consideration of the development plan and perhaps discussion with the planning authority, as well as recognition of any other legal constraints on the site such as covenants, rights of way etc. Under the US system this represents the legal feasibility evaluation stage. Next we narrow our choice further by considering financial feasibility. At this point we would need to examine the macro economic climate and local market and sub-market indicators, taking into account interest rates, the property cycle, current and projected supply and demand and local market conditions. The calculation would also involve consideration of development costs, including site acquisition, clearance, construction and/or refurbishment set against the development value of the completed scheme. Only those possible uses which fall within the bottom right hand box would therefore be worthy of detailed consideration. In this way appraisers and development consultants can demonstrate to their client that they have taken a rational approach and considered all the angles. It is of course what many practitioners/developers do unconsciously but we would advocate an explicit approach using the above methodology.

The residual method

In the market for property with development potential an adaptation of the income approach, known as the 'residual method', is the most widely used technique.

The conventional approach to a residual valuation is based upon a very simple concept. The value of the completed project less all of the costs incurred in the project is equal to the residual value, or:

Gross development value   less   All costs and profit   =   Site Value

The value of the completed project is usually known as the Gross Development Value. This is the capital value of the finished development and is usually found by capitalisation of the projected income from the scheme.

The costs of development include all of the costs of construction, including professional fees (architects, quantity surveyors, etc), the cost of borrowing the capital to undertake the scheme, and the developer's profit. The surplus that remains after deducting all of these costs from the projected value is the residual value of the bare site or site and buildings to be redeveloped.

Exploring the main inputs to a residual appraisal

It has already been suggested that one of the difficulties in applying the residual method revolves around the number and variability of the inputs.

Completed development value

This is the market value of the completed development which would be arrived at by the comparison or investment method as appropriate, depending upon the type of property.

In the case of commercial developments, it is assumed that on completion, the developer will sell the development to an investor. This will incur agents fees, advertising and legal fees which may amount to approximately 3% of the value of the completed development. In addition it will be easier to sell if the building is already occupied and so letting agent's fees, which might account for some 10% of the first year's rent, need also to be taken into account.

Rent and yield

In the case of most commercial/industrial buildings, the completed development value will be calculated using the investment method of valuation. This will require an estimation of the rent of the development and the likely investment yield. Both will be based on current market evidence. The rent is normally based on the lettable area of the completed development. This will usually be the Net Internal Area. It should be noted that in a residual appraisal it is the current rent which is used notwithstanding the fact that rental levels may have changed by the time the development is completed. It would be extremely dangerous to base an appraisal on inflated future rents as there can be no guarantee that these will be achieved. The possibility of rental increase/decrease over the period of the development is something which can be reflected in the risk or profit allowance.

Incidental costs

- It is assumed that the developer will dispose of the completed development and will thus incur incidental disposal costs which will include agent and legal fees amounting to around 2.5%.
- The developer will also incur costs in acquiring the site including land value stamp duty at 1–4%, legal fees of between 0.25 and 0.5% and agents fees of 1–2%.

Costs

The main costs to be deducted from the completed development value will include pre and post development costs. The pre development costs are those occurring before the commencement of construction:

- demolition
- infrastructure, planning gain and compensation
- site investigation
- ground investigation
- land surveys
- planning fees
- planning consultant fees
- legal costs for appeals
- building regulations, scale fee based on final building cost
- funding fees.

Construction costs

Construction costs are usually assessed at a price per m². Sources of cost information include Spons Architects and Builders price book which is published annually. In practice the actual construction costs will normally be based on the quantity surveyor's estimate.

Fees

The professional fees for architects, QS, Engineers etc need to be added to the construction costs. It is usual to add 12–13% to the construction costs but this will be dependent upon the nature and scale of the development. In more complex schemes it may be necessary to make additional allowance for project manager's fees.

Finance

- The developer will need to take out short term loans to fund the development. Even if the developer is able to fund the development from capital, this will incur opportunity costs. Funding needs to cover the length of the development period which must reflect the pre contract period at the outset and any letting or void period once the development is completed. It is only on successful completion and sale that the developer will be able to recoup costs.

- The basic residual usually adopts a 'rule of thumb' approach to the calculation of finance charges by either taking half the development period or half the building costs. This is usually adopted to reflect the average cost. Clearly the developer will not borrow the whole of the cost of the development at the beginning of the development period as payments to contractors will be staged as the development progresses,
- The costs of short term finance can be anything between 1 and 6% above the prevailing base rate. This will depend upon the status and track record of the developer as well as the type, quality and location of the development.
- The developer will also need to allow for the cost of interest payments on the site acquisition price. Where a site is being purchased at the outset, the cost of this must be funded over the whole period of development, from purchase through to disposal.

Marketing/letting promotion

It is necessary to allow for both the letting and sale of the completed development. This will include:

- advertising
- particulars and brochures
- site boards
- mail shots
- launch ceremonies
- show suites
- public relations
- letting incentives.

Miscellaneous costs

Other costs which need to be reflected may include:

- party wall agreements
- rights of light agreements
- void periods — maintenance and insurance
- empty rates.

Some costs are difficult to identify or estimate accurately and these are sometimes reflected in a contingency of around 3–5%. It is sometimes argued that these elements are included in the profit/risk element. Thus, in the case of a development with major cost uncertainty the developer will seek to make a higher allowance for profit.

Developers profit or risk

If no allowance was made for profit, the developer would be undertaking the development for nothing. This allowance is normally related either to the cost of the development or the completed development value; typically this might be 20–25% cost, or 15–20% value.

But profit is a major variable, dependent on many factors. If the market is competitive, developers may be forced to trim profit margins, however if the risks are high, for example if the time taken to let or sell the development is extended in a weak market, then the profit element would need to be increased.

Example 11.1

A plot of land has planning permission for the erection of 7,000 m² (gross) of office space on five floors.

The development will be completed in two years from now and rents are expected to average £120 per m² (net). Building costs are expected to average £400 per m², excluding fees.

Prepare a valuation to advise a prospective developer of the maximum bid to be made for the site. (Comparable evidence of prices obtained for similar sites is not available.)

Gross development value:

7,000 m² (gross) × 85%*	=	(say) 6,000 m² (net)
6,000 m² @ £120 per m² pa	=	£720,000 pa
× YP perp @ 7%		14.29
		£10,285,714

*Reduction for non-usable space of 15%.

Gross development value (GDV) c/o : £10,285,714

Less costs:

(a) Building costs:
7,000 m² @ £400 per m² £2,800,000

(b) Architects and quantity surveyors fees @ 12.5% of building costs £350,000
Total building costs: £3,150,000

(c) Finance costs:
15% pa for 2 years on 50% of total building costs (£1,575,000 × (1.15)²)
− £1,575,000 = £507,937

(d) Legal fees, estate agents' fees and advertising upon disposal:
@ 2% GDV = £205,714

(e) Promotion say: £50,000

(f) Developer's profit: @ 15% GDV £1,542,857

Total costs: £5,456,508
Residual: (GDV − costs) £4,829,206
Let site value = x
Legal and valuation fees on site purchase @ 3% = 0.03x
Total accumulated debt after 2 years @ 15% pa = $1.03x(1.15)^2$
$1.03x(1.15)^2$ = £4,829,207
1.362x = £4,829,207
Site value = £3,545,673

The use of the residual method varies according to the stage of the development process. The above calculation is very generalised in its assumptions and the technique is often criticised as a method of determining land value.

Where it is used to assess the value of bare sites for rating or statutory purposes it is often divorced from the realities of actual development and has been criticised by the Lands Tribunal, whose members denounced the method as 'far from a certain guide to values' (*Cuckmere Brick Co Ltd and Fawke* v *Mutual Finance Ltd* (1971) 218 EG 1571) and have suggested that 'once valuers are let loose upon residual valuations, however honest the valuers and however reasoned their arguments, they can prove almost anything' (*First Garden City Ltd* v *Letchworth Garden City Corporation* (1986) 200 EG 123).

However, as will be explained later in this chapter, where the technique is used to appraise projects where land has been purchased for a known sum and a quantity surveyor has provided accurate cost information the technique is more robust.

Example 11.1 can be used to illustrate the dangers of the generalised assumptions built into the simple residual valuation.

For example the finance for development is calculated on only 50% of the total construction costs to reflect the phased drawing down of borrowing over the development period. This assumption will hold true if the payments to contractors follow an equal distribution pattern similar to that shown in Figure 11.2 for the traditional office building. However, if the building involved a portal steel frame which brings the development costs forward, the expenditure pattern may change to that shown in Figure 11.2 for an industrial building. This destroys the 50% assumption and an expenditure of this pattern should assume funding on around 75% of the total building costs.

Figure 11.2

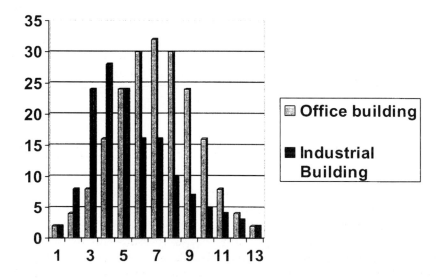

The need for accurate cash flow based projections of building costs are therefore essential for accurate residual valuations.

In addition to being subject to generalised assumptions the final value is very sensitive to small variations in key variables — particularly the yield, rental values and building costs.

To illustrate this problem note the effects on the residual value calculation in example 11.1 with the changes to the key variables as shown in Figure 11.3.

Figure 11.3 Illustration of the sensitivity of the residual site value calculated in Example 11.1 to changes in key variables

Original site value	£3,545,673
Site value when yield only increased by 0.5%	£3,120,825
Site value when rents only increased by 10%	£4,184,426
Site value when building costs only increased by 10%	£3,280,977

A further problem facing valuers has been the difficulty of handling an increasing number of variables such as VAT, fees, etc in the valuation exercise when carried out manually. Currently many development surveyors use spreadsheets or computer software packages which do not remove the basic criticism but their flexibility and speed of operation enables valuers to check each appraisal for sensitivity and avoid generalisations in the assumptions to achieve greater confidence in their opinions of value. (An example of this approach is provided in Appendix C.)

Example 11.1 has been reappraised using a package prepared and marketed by Circle Systems which is one of several programs now available to development valuers.

The report produced by the system is set out in Appendix C, together with notes on the assumptions and methodologies used.

The traditional residual approach can therefore be significantly improved with the application of computers through sensitivity analysis and more complex treatment of the underlying assumptions.

However, although packages such as Circle allow for complex phasing of developments by the linking of parts and areas of schemes together they still rely heavily on assumptions made about the distribution of payments and receipts.

The need for adoption of the cash flow approach

Cash flow modelling is now a standard feature of most development appraisal systems and its use is encouraged to counter the difficulties and criticisms described above.

These are especially useful when dealing with large phased developments such as business parks or major residential schemes. The cash flow approach also allows the valuer to take into account the financial status of the client and to provide a more accurate and realistic financial appraisal. For example, joint ventures or schemes involving equity finance may require differential interest rates and complex flows of income which cannot be accommodated in a traditional appraisal.

As the valuation profession is being encouraged to be more conscious of the needs of its clients the cash flow approach should be more widely adopted to ensure that the financial situation of the client is accurately modelled within the appraisal.

The cash flow approach explained

Testing the traditional residual results against cash flow models suggests that the greater variability occurs with phased developments such as low-rise residential schemes where cash in-flows coincide with cash out-flows. The variation in result is less pronounced with the relatively simple one-off development, especially those with a project term of 12 months or less.

In the absence of alternative development schemes, it is almost certain that the cost of borrowing money will be at least as high as the rate of interest that the developer could earn on excess funds. It is therefore reasonable to conclude that the developer should pay off debts as soon as income is produced by the development.

The developer will pay considerable attention to his likely cash flow and his likely maximum borrowing requirement over the building period. A cash-flow table will provide full information concerning these two points, and will allow for the accurate estimation of finance charges. In very complex multiuse schemes the development surveyor and quantity surveyors may be able to use cash flow analysis to reschedule the detailed timing of development to improve the overall return on capital.

Systems such as Circle allow the user to construct a monthly cash flow from the original residual valuation which can then be edited to reflect more accurate data provided by quantity surveyors or to explore the effects of adjusting the cash flow. The program provides the user with a full DCF analysis of the project and can also be converted into a Lotus 1-2-3 spreadsheet file.

The example below illustrates the cash flow approach using a simple manual calculation.

Example 11.2

A site has the benefit of planning permission for the erection of a block of 20 apartments, totalling 1,100 m^2 of space.

The value of each completed apartment is estimated to be £100,000, using comparable transactions, but flat prices are expected to continue to rise at 5% pa.

Building costs are estimated at £800 per m^2 and are expected to rise at 0.5% per month. It is expected that 5 apartments will be sold in each of months 9–12 and that building costs will be evenly distributed over a 12-month building period.

Architects' and quantity surveyors' fees will be payable in two instalments in months 6 and 12 at 10% of the building costs.

Agents' and solicitors' fees will be charged on sale at 1% of the sale price of each apartment, and the developer will require a profit of 15% of the sale price.

Advise the developer as to the maximum bid for the site.

1   Apartment prices are increasing at 5% pa or 0.4% per month.

In month 9,
5 flats will sell for £103,660 each   =   £518,300

In month 10,
5 flats will sell for £104,070 each   =   £520,350

In month 11,
5 flats will sell for £104,500 each   =   £522,500

In month 12,
5 flats will sell for £104,910 each   =   £524,550

2   Building costs are estimated at £800 per m2, spread evenly over 12 months, rising at 0.5% per month.

Total building costs for the entire block of £880,000 divided by 12 gives a cost of £73,330 in month 1 which will increase by 0.5% in month 2, etc.

This results in a building cost flow of :

month 1:	**£73,330**
month 2:	**£73,700**
month 3:	**£74,060** — *mistake*
month 4:	**£74,440**
month 5:	**£74,810**

in month 6 fees are payable on the costs to date at a rate of 10%
(10% of £445,490 = £44,549)   £75,180 + £44,549 =

month 6:	**£119,729**
month 7:	**£75,560**
month 8:	**£75,940**

in month 9 the first of the sales of 5 flats is achieved adding fees @ 1%
and profit @15% of the sale price to the costs
(1% of £518,300   =   £5,183 and
15% of £518,300   =   £77,745)

£76,320 + £5,183 + £77,745   **month 9:   £159,248**

in month 10:   £76,700 + £5,204 + £78,053
(calculated as above but on
month 10 sale prices)   **month 10: £159957**

in month 11:   £77,080 + £5,225 + £78,375   **month 11: £160680**
                          (see above)

in month 12:   £77,470 + £5,246 + £78,683   **month 12: £161399**
                          + £45,900 (Arch/QS fees @
                          10% costs months 7–12)

From this information a cash-flow table can be constructed incorporating the effects of interest charges at 1% per month.

Month	Benefits(£)	Costs (£)	Net (£)	Capital (£) *(Total Capital)*	Interest (£) Outstanding
1	–	73,330	−73,330	−73,330	733
2	–	73,700	−73,700	−147,760[1]	1,480
3	–	74,060	−74,060	−223,301	2,233
4	–	74,440	−74,440	−299,974	3,000
5	–	74,810	−74,810	−377,784	3,778
6	–	119,729	−119,729	−501,291	5,013
7	–	75,560	−75,560	−581,863	5,819
8	–	75,940	−75,940	−663,622	6,636
9	518,300	159,248	+359,052	−311,206	3,112
10	520,350	159,957	+360,393	+46,075[2]	+ 461
11	522,500	160,680	+361,820	+408,355	+4,084
12	524,450	207,299	+317,251	+729,690[3]	

Residual	=	£729,690
Let site value	=	x
let fees on site purchase	=	3% = 0.03x

Then

1.03x + interest for 12 months at 1% per month	=	£729,690
1.03x(1.01) 12	=	£729,690
1.03x(1.1268)	=	£729,690
1.1606x	=	£729,690
x	=	£628,718
Site value	=	£628,718

Notes:
1. Already outstanding from month 1 is a debt of £73,330, and an interest charge of £733. Added to these is a new loan of £73,700, the total being £147,763 on which the interest charge for month 2 will be levied.
2. This figure if positive, as the total debt of £311,206 plus £3,112 interest is more than repaid by the receipt in month 10 of £360,393. Hence a surplus of £46,075 remains. This can be invested to earn interest at 1% per month.
3. This final surplus remains after paying all costs except the cost of the site itself, and fees and finance on its purchase. Calculated as before, the developer's maximum site bid is £628,718.

Example 11.3

In Example 11.2 the cash flow has been undertaken on a terminal value basis.
In this example the same development is valued using a present value approach.

Month	Benefits(£)	Costs (£)	Net (£)	PV£1 @ 1%	PV Outstanding
1	–	73,330	–73,330	0.99001	–72,604
2	–	73,700	–73,700	0.98030	–72,248
3	–	74,060	–74,060	0.97060	–71,882
4	–	74,440	–74,440	0.96098	–71,535
5	–	74,810	–74,810	0.95147	–71,179
6	–	119,729	–119,729	0.94205	–112,270
7	–	75,560	–75,560	0.93272	–70,476
8	–	75,940	–75,940	0.92348	–70,129
9	518,300	159,248	+359,052	0.91434	+328,296
10	520,350	159,957	+360,393	0.90529	+326,259
11	522,500	160,680	+361,820	0.89632	+324,308
12	524,450	207,299	+317,251	0.88745	+281,544

NPV = £647,563

Let site value	=	x
let fees on site purchase	=	3% = 0.03x.

Then

1.03x	=	£647,563
x	=	£628,702 (variation from example 11.2 due to rounding).
Site value	=	£628,702

An important prerequisite for a cash flow approach is an accurate scheduling of construction activities. This uses network analysis, or critical path analysis whereby critical sequential events become the critical path and other non-sequential events can be scheduled to run in parallel; in turn some of the parallel costs may also be sequential.

The end result is the production of a project network with a start and finish date with all the intermediate critical dates. The actual costs can then be estimated and transferred to a cash flow. The two together will then become part of the project manager's management tools for monitoring the development programme. Such an approach, if applied to example 11.1, would highlight the need to reduce the construction period in order to reduce the high finance costs. The need to reduce finance costs in periods of high interest rates has played its part in developing innovative 'fast track' construction techniques.

Cash flows and residuals can be calculated on a current cost basis, or on a future cost basis building in variations in rents as well as variations in labour and materials. None of the computerised sophistications can, however, overcome the problem that the acceptability of residual and cash flow development appraisal methods rests not with their rationale, which is irrefutable, but with the quality of the evidence used by the appraisal team to estimate costs and benefits.

Viability studies

Early in this chapter it was stated that the residual method of valuation is based upon a simple equation:

Gross development value − development costs (including profit) = Residual value

The aim of the residual valuation is to find the unknown in this equation, the residual value. Often, however, a prospective developer will be aware of the likely site cost and consequently of the cost of fees and finance, either because the vendor has stated an asking price, or because negotiations have revealed the minimum figure which the vendor will accept. This is particularly likely in the case of an existing building which is to be improved or replaced.

In such a case the single unknown in the equation can now become the developer's profit. The aim of viability study is to assess the profit likely to be made from a development scheme, given the cost or asking price of the subject site.

Computer packages, for example Circle, allow the user to select from a drop down menu whether the system is to calculate residual land value or developer's profit. The input screens and processing are then adjusted according to the user's selection.

The cash flow approach is advocated by most software manufacturers, if not valuers, because of its greater accuracy and ability to model the payment of contractors, architects and finance charges much closer to the actual payment profile than the smoothed assumptions often contained in a traditional appraisal.

In addition a detailed cash flow approach allows the modelling of complex financial arrangements that may often underpin major property developments around the world. For example, equity sharing and joint ventures where the interest rates may differ for each project partner depending upon their risk profile.

The cash-flow table is particularly adaptable as a viability study and the authors recommend that valuers using appraisal software always utilise a cash flow approach and only use the initial appraisal as a way of inputting data and obtaining a summary picture of the development analysis.

Example 11.4

Using the facts of example 11.2, advise the developer of the potential profit if the total cost of the site, including fees, is £500,000

Month	Benefits(£)	Costs (£)	Net (£)	Capital (£)	Interest (£) Outstanding
1	–	573,330	–573,330	–573,333	5,733
2	–	73,700	–73,700	–652,763	6,528
3	–	74,060	–74,060	–733,351	7,334
4	–	74,440	–74,440	–815,124	8,151
5	–	74,810	–74,810	–898,086	8,981
6	–	119,729	–119,729	–1,026,796	10,013
7	–	75,560	–75,560	–1,112,623	11,126
8	–	75,940	–75,940	–1,199,690	11,997
9	518,300	81,503	+436,797	–774,890	7,749
10	520,350	81,904	+438,446	–344,193	3,442
11	522,500	82,305	+440,195	+92,561	+926
12	524,450	128,616	+395,934	+489,420	

£489,420 represents the developer's profit

This can be expressed as $\dfrac{£489,420}{£2,085,700}$ = 23.47% of GDV

or as $\dfrac{£489,420}{£1,515,897}$ = 32.29% of total costs

The residual method has become a straightforward DCF exercise. Nevertheless, there will still be areas of variability, and the criticism that the result is sensitive to changes in inputs remains. The next stage is therefore to incorporate probability measures in the analysis (see Chapter 13).

Rents, costs and interest charges may be weighted according to the valuer's expectation of possible changes, such an approach being eminently suited to computer analysis.

Question 1

Assess the residual site value of a parcel of land suitable for 5,000m² of office space.
The scheme will take 12 months to complete. Rents are £200 per m² (net lettable area).
Construction costs are £400 per m².
The scheme would sell on a 7% basis and finance is available at 14%.
Promotion costs are budgeted at £40,000.
Letting fees are agreed at 15% of the first year's rent.
Purchases costs are 1.75%,
site purchase costs 1.5%, stamp duty 1%.
Agents' fees are 1% of GDV.
Developer's profit required is 15% of GDV.

Summary

- Residual valuations or development appraisals are used to assess the value of land and buildings with latent development potential.
- The method is based upon established economic principles and the method itself is logical.
- However, the accuracy of the residual method relies on the valuer's ability to estimate all of the variables within the calculation. The costs and benefits must be carefully estimated with respect to the development market, using hard evidence where possible to support the rents, capitalisation rates and building costs adopted.
- Development appraisal is an ideal application for information technology, allowing the valuer to assess the sensitivity of the valuation to changes in the underlying variables.
- The final opinion of development value should, wherever possible, be checked against comparable market evidence. However, as each development opportunity is unique in terms of its highest and best use, market comparison will only provide the valuer with a basic 'yardstick' with which to cross reference the appraisal.

Spreadsheet User 11

Using Excel to build an appraisal

Value

We will now explore a simple development appraisal using a cash flow approach by building an Excel spreadsheet to undertake the development appraisal.

Appraisal data

Construction of 2000m² (lettable area) offices
Estimated rental value per m² on completion £1000 paid in advance

Estimated time from completion of construction 12 months to full occupation.

Estimated take up rate: 500m²/quarter
Construction complete   12 months time
Construction commences   immediately

Using this data we can begin to construct a development appraisal spreadsheet.

(i)   Enter quarterly time periods across the spreadsheet

     0   1   2   3   4   5   6   7   8

(ii) Enter the rental income commencing in quarter 4 (remember that in Excel payments are treated as occurring at the end of the period when using NPV and IRR functions)

(iii) Assuming that an investor has made an agreement to purchase the development when it is fully let at an agreed all risks yield of 8%.

We have now completed the (positive) value inputs into the development and your spreadsheet should be similar to the following:

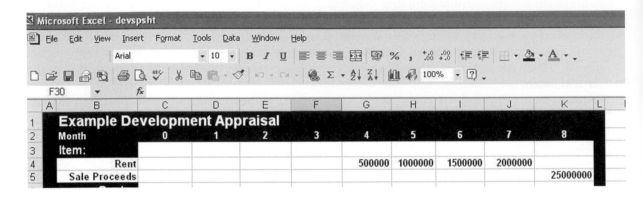

Note: The sale proceeds have been calculated by applying the all risks yield to the total income when the development is fully let.

ie   2000m² @ £1000/m²   =   2,000,000

$$\times \text{ Years Purchase} \quad \left(\frac{1}{\text{yield}}\right) \quad = \quad \frac{1}{0.08} \quad = \quad \times 12.5$$

=   £25,000,000

We have now considered the value side of the residual equation by considering the income flow during the development period. Now we need to consider the costs of development.

Estimating the costs of development

The costs of development can be broken down into a series of categories.

Pre-development costs

These are the costs that occur before the construction commences and in some cases even before the site is acquired.

Examples
1. Planning permission fees charged by local authorities.
2. Building control fees charged by local authorities.
3. Consultants fees — planning consultants to assess the likelihood of achieving the necessary consent; engineers to assess the site characteristics for foundations.

4. Infrastructure: roads, utilities — gas, electricity, water, drainage and sewerage.
5. Planning 'gain': contributions to the local authority to obtain planning permission. In the UK usually in the form of amenity improvements such as landscaping or road improvement. But in North America may also involve cash, density transfer or building of social housing,
6. Compensation: where there are existing buildings on a site to be demolished and tenants have security of tenure compensation may have to be paid.

Construction costs

The costs incurred in the physical construction of the development.

Examples

1. Demolition — of any existing structures.
2. Construction — usually worked out for the initial appraisal on a figure per square metre basis provided by tables produced by builders/architects or professional bodies. Note: These rates are applied to the external envelope of the proposed buildings. When the development reaches a more advanced stage a quantity surveyor will provide more accurate month by month construction cost estimates.
3. Contingency — frequently a percentage of the building costs (typically 5%–6%) is built in for unforeseen contingencies that frequently arise during building projects.
4. Professional fees — these are normally calculated as a percentage of the building costs, and include the architect, the quantity surveyor, the structural engineer, the mechanical and electrical engineer, and the project manager, if applicable. They are normally 12–13% of the building cost and are based on the scale charges of each profession, a negotiated percentage or a fixed fee. The percentage agreed with each member of the professional team depends on the nature and scale of the development. Small refurbishment schemes normally attract higher percentages than larger, complex development projects. If a developer will need to appoint other professionals, such as a traffic engineer, a landscape architect or a party wall surveyor, then these need to be included in any evaluation of the project.
5. VAT — this may be applicable to both the construction costs and the professional fees depending upon the status of the parties and the VAT (or sales tax) regulations of a particular country.

Finance

The cost of finance in a cash flow approach will be taken care of in the discounting provisions of the calculation of the NPV and IRR calculations. However, where a non-cash flow simple appraisal is undertaken the finance must be calculated using assumptions as to the timing of payments.

Marketing/letting and promotion

Examples

1. Agents letting — frequently a property agent will be used to let the completed property and they will charge a percentage (typically 7%–10%) of the first years rent. These fees may be subject to VAT.

2. Marketing/promotion — a variety of costs may be incurred which are in addition to the agents fee, eg:
    – Advertising — newspapers, journals, Internet, radio/TV
    – particulars and brochures
    – site boards/hoardings
    – direct mail shots
    – launch ceremonies
    – show suites/offices
    – public relations
3. Letting incentives — in order to secure a letting, incentives may be offered which either incur:
    – a real cost — a cash contribution to the tenant for fitting out
    – an opportunity cost — a rent free period where potential income is foregone
    These must be built in as a cost in the appraisal.
4. Funding fees — the developer may have to pay a fee to a specialist financier for both the arrangement of complex funding and to a solicitor for drawing up the appropriate documentation.

Site costs

1. The actual price paid for the land.
2. Legal costs associated with the purchase.
3. Agents costs associated with the purchase.
4. Stamp Duty/property tax.
5. VAT (or sales tax) on any of the above.

Building the costs into our Excel spreadsheet

Returning to our spreadsheet we must now enter the following costs into our simple appraisal:

Pre-development costs (all occurring in the first month)

1.	A planning fee of	£10,000
2.	A building permit fee of	£5,000
3.	Site investigation	£20,000

Construction costs

1.	Construction costs	£2,000,000 end of quarter 0
		£6,000,000 end of quarter 1
		£6,000,000 end of quarter 2
		£2,000,000 end of quarter 3
2.	Contingency	5% of construction costs
3.	Professional fees	10% of construction costs (excluding contingency)

Finance
Finance is available at an interest rate (cost of capital) of 2.5% per quarter.

Marketing
Agents letting fees: 7.5% of initial rent.

Promotion

£50,000	Quarter 0
£20,000	Quarter 1
£10,000	Quarter 2
£10,000	Quarter 3
£10,000	Quarter 4
£20,000	Quarter 5

Site costs (All payable in Quarter 0)

1.	The site costs	£4,500,000
2.	Agents and solicitors fees amount to	£100,000
3.	Property tax	£50,000

When you have entered the data into the spreadsheet it should look something like this:

	Month	0	1	2	3	4	5	6	7	8
	Example Development Appraisal									
3	Item:									
4	Rent					500000	1000000	1500000	2000000	
5	Sale Proceeds									25000000
6	**Costs:**									
7	Planning Fee	-10000								
8	Builidng Fee	-5000								
9	Site Invest.	-20000								
10	Construction	-2000000	-6000000	-6000000	-2000000					
11	Contingency	-100000	-300000	-300000	-100000					
12	Prof. Fees	-200000	-600000	-600000	-200000					
13	Letting Fee					-37500	-37500	-37500	-37500	
14	Promotion	-50000	-20000	-10000	-10000	-10000	-20000			
15	Land Cost	-4500000								
16	Fees	-100000								
17	Tax	-50000								

Now we have to calculate the cash flow.

Use the Excel Sum function to add the figures in each column.
In our spreadsheet in cell C19 type (or use Autosum) = Sum (C4:C17).

This can then be copied across to all the rows for Quarter 0 to Quarter 8 using the copy and paste buttons/keys.

You now have the undiscounted cash flow.

Calculating the NPV

Next:

Enter the formula for the Present Value (PV) $\quad = \quad \dfrac{1}{(1 + i)^{n}}$

into the cell below the first cash flow figure (in our spreadsheet C20).

Remember you will need appropriate brackets and that we are using quarterly periods and a quarterly interest rate.

The contents of our cell C20 are: $(1/((1 + 0.025)^{\wedge} C2))*C19$

This formula can then be copied across all the quarters.

You can then use the sum function to add all of the quarterly discounted cash flow figures to give you the Net Present Value (NPV).

Alternatively, you can use the Excel NPV function by typing in:

= NPV (discount rate, range of cash flow)

which in our case = NPV (0.025, C19:K19)

If you use the above you will note that it does not give the same figure as calculating it using the PV formula and summing the discounted cash flows. This is because the function assumes the investment begins one period before the date of the first value and that the initial payment is made at the end of the first time period. In fact in our appraisal as in most property cash flows the site is purchased at the start of the first period.

Therefore, the NPV function must be modified in our case to:

= NPV (0.025, D19:K19) + C19

This function should now give you the 2165410 expected result.

You are advised to print out, read and file in this section the Excel helpsheet on the NPV function.

The Net Present Value represents the developers profit and can be converted to a return on capital costs or value.

Calculating the IRR

The Internal Rate of Return (IRR) can now be calculated using the Excel IRR function:

= IRR (range of cash flow, trial rate)

In our spreadsheet we have entered into cell C22 the formula:

= IRR(C19:K19, 0.05)

= 3.960%

Remember a trial rate may be needed (we have used 0.05) to initialise the process of iteration.

You are advised to print out, read and file in this section the Excel helpsheet on the IRR function.

When you have calculated the NPV and IRR your spreadsheet should look something like this:

	Microsoft Excel - devspsht

File   Edit   View   Insert   Format   Tools   Data   Window   Help

Arial   ▾ 10 ▾   **B** *I* U   ≡ ≡ ≡ 🖾 💲 % , ‰ ‰ 津 津 🖾 ▾ 🖎 ▾ A ▾

F30   ▾   *fx*

	A	B	C	D	E	F	G	H	I	J	K	L
1		**Example Development Appraisal**										
2		Month	0	1	2	3	4	5	6	7	8	
3		Item:										
4		Rent					500000	1000000	1500000	2000000		
5		Sale Proceeds									25000000	
6		Costs:										
7		Planning Fee	-10000									
8		Builidng Fee	-5000									
9		Site Invest.	-20000									
10		Construction	-2000000	-6000000	-6000000	-2000000						
11		Contingency	-100000	-300000	-300000	-100000						
12		Prof. Fees	-200000	-600000	-600000	-200000						
13		Letting Fee					-37500	-37500	-37500	-37500		
14		Promotion	-50000	-20000	-10000	-10000	-10000	-20000				
15		Land Cost	-4500000									
16		Fees	-100000									
17		Tax	-50000									
18												
19		CASH FLOW	-7035000	-6920000	-6910000	-2310000	452500	942500	1462500	1962500	25000000	
20		P.V at 2.5%	-7035000	-6751220	-6577037	-2145065	409943	833033	1261109	1650983	20518664	
21		N.P.V =	2165410									
22		I.R.R =	3.960%			Effective annual rate :			16.806			
23												
24												

Remember, this is a quarterly IRR and to convert to an effective annual rate you must use the formula: $(1 + i)^{nth} - 1 \times 100$ where nth is the number of periods pa.

Therefore, in our example:

$(1 + 0.03960)^4 - 1 \times 100$ (4 = quarterly periods)

= 16.806%

In our spreadsheet we have entered into cell I22 the formula (((1+(C22))^4)*100.
If your cell C22 is not in a percentage format your formula must read: (((1+(C22/100))^4)–1)*100.

The Profits Method

Introduction

Where comparable rental transactions are not available, it may be necessary to adopt a profits method of valuation based on the audited accounts of a business. The rationale for the approach is based on the economic theory that the rental value of a property is a function of its earning capacity, productivity or profitability.

For most commercial and industrial properties comparable transactions are available and rental values can be ascertained by analysing recent market behaviour. Comparables will not normally be available when a property enjoys an element of monopoly. A monopoly may be either factual or legal. Property with an element of factual monopoly might include a seaside car park, a railway station kiosk, a racecourse and an amusement park. In these cases little, if any, competition will exist. A legal monopoly, however, may arise where a business can only operate with a licence and where perhaps only one licence is issued for a large area — this could be the case with a licence to sell alcohol.

In addition, the profits method may be used to value a business property where accounts are available and where the comparative method cannot be employed easily because of a lack of satisfactory evidence. Examples might include hotels, theatres and cinemas.

In outline, the profits (or accounts) method of valuation requires an estimate of the gross earnings that can be earned by the business in the property. All normal working expenses are then deducted excluding any rent or loan interest payments on the property. The resulting figure, known as the 'divisible balance', represents the amount available for the tenant's share (or remuneration) and the landlord's share. Thus, a deduction of the tenant's share leaves the surplus for payment to the landlord in the form of an annual rent.

Example 12.1

A simplified model of the profits method of valuation takes the following form:

(a)	Gross profit		
	Estimated Gross Earnings	£220,000	
	Less purchase of goods	£120,000	
			£100,000

(b) Less
     Working Expenses
     (Excluding Rent)   £70,000

     Divisible balance   £30,000

(c) Less
     Tenant's Share:
     Interest on Capital   £5,000
     Remuneration   £15,000
                                                                         £20,000

(d) Surplus available as rent (payable to landlord) = £10,000 a year

This rental, derived from accounts, can then be capitalised to arrive at an opinion of market value.

Notes

(a) Gross profit

The valuer will normally examine the last complete set of accounts produced, before the date of valuation, together with the preceding two or three years' accounts. The aim is to determine a level of profit that can reasonably be expected in the future, reflecting any past trends or fluctuations in the profit. The actual accounts may not always be a good indicator of the earning capacity of the property as, for example, the current business may have been operated by either a workaholic or a sloathful character lacking in business acumen. The aim should be to determine the profits which a tenant of reasonable competence could make from occupation of the property.

(b) Working expenses

These will include such items as wages and insurance, heat and light, telephone, cleaning, advertising, printing, postage and stationery, insurance, accountancy charges and depreciation of fixtures, fittings and equipment. As the final balance is to represent the surplus available as rental value for the property, any rent that is actually paid is not deducted as a working expense. This includes any ground rental payment and/or any interest payable by the operator on loans.

(c) Tenant's share

The deduction for the tenant's share represents:

*   Interest on the tenant's own capital tied up in the business; that is, capital for the purchase of fixtures and fittings, equipment furnishings, stock, cash, etc and
*   Remuneration for the tenant's time and effort in running the business. The term 'tenant' in this context may be the owner occupier. The deduction represents the 'opportunity cost' of capital and labour (very simply, salary or wages).

(d) Rent as a surplus

The method of valuation illustrates that the rental value of all land and property is derived from earning capacity or productivity. The valuer is required to calculate rental value from the viewpoint of a potential occupier of the property. In contrast, the comparative method merely requires the valuer to study the outcome of negotiated transactions between several landlords and tenants. Nevertheless, it must be remembered that the tenants in these cases will probably have prepared a profits type of calculation to convince themselves that the rent agreed can be paid out of the surplus earnings of the proposed business.

The RICS Guidance Note 1 *Trade-related valuations and Goodwill* confirms these general principles, in particular it states that:

> ... The task of the valuer is to assess the fair maintainable level of trade and future profitability that could be achieved by an operator of the business upon which a potential purchaser would be likely to base an offer. When assessing future trading poterntial the valuer should exclude any turnover and profit that is attributable solely to the personal skill, expertise reputation and/or brand name of the existing owner or management. However, in contrast , the valuer should include any additional trading potential which might be realized under the management of an average competent operator taking over the existing business at the date of valuation ...

The British Association of Hotel Accountants (1993) confirms these general notes of guidance in relation to hotels

> Hotels should be valued by reference to their recent/current performance and future trading potential. This is achieved by the 'Income Capitalisation' approach which seeks to assess value by reference to projected net cash flows. It requires the application of a capitalisation factor which may be either a multiple of marketable earnings or a Discount Rate.

There has been strong criticism of the BAHA view on the use of a full DCF approach from certain sections of the leisure industry. It is argued that DCF is a technique for assessing worth to a specific individual rather than for assessing market value and that the proper approach is the simple process of capitalising average or 'maintainable' earnings. DCF is considered to be more appropriate for appraisal or analysis of development projects rather than existing operational entities. As is often the case in discussion on valuation methods all opinions have their strengths and weaknesses.

The appropriateness of the use of the two alternatives in the specialist area of valuation of leisure and other businesses sold as operational entities might be summarised as follows:

- analyse the market and the behaviour of buyers and sellers in the market
- identify the most probable type/class of purchaser having regard to property size and use
- adopt the appropriate market valuation technique.

In practice smaller properties in this category, such as small family hotels, retirement homes, pubs, might be valued by a form of direct comparison or quasi-profits approach where there is strong market evidence. Analysis of the relationship between expected maintainable earnings and MV, price, provides a realistic view of the probability of a sale being achieved at the figure obtained, simply by relating the absolute return and rate of return to the level of return a rational person would consider to be adequate from such an expenditure.

In the case of larger properties which would attract core operators in the sector the valuer will need to assess the level of maintainable profit, as previously described, and to apply a multiplier based on the valuer's knowledge and experience of the marketplace.

In the case of substantive properties or new developments, where there will be less direct evidence in the market and where purchasers and investors are driven by current and future profitability, there is a strong case for a full DCF valuation.

The strongest case for DCF seems to lie where a new enterprise is planned and where there is a need for a viability study. The crux of the problem here is that new schemes generally take a number of years to mature to their full potential. A parallel can be drawn here with, say, a large residential development or a major retail scheme where sales and/or lettings will take a number of years to complete. The result is a fluctuating cash flow which can be realistically estimated by the key players based on their experience of similar schemes in other localities. The valuer, in seeking to mirror the market behaviour, should therefore follow market practice.

The difficulty is that having accepted the argument for using a form of projected cash flow and discounting techniques for assessing MV for substantive new schemes there seems to be no logical argument for rejecting its use in circumstances where such cash flows should be capable of more accurate measurement — that is, where a maintainable business is supported by current accounts.

The argument for the use of the simpler 'income capitalisation' is that it is derived direct from the market but, as has been seen in the case of investment property, this approach is an implied approach where increasingly lenders and purchasers are seeking a more explicit assessment of the return to be achieved in the short to medium term.

> The DCF approach is more rigorous in that it requires consideration to be given to the dynamic nature of the future income of the hotel, inflation and interest rates. However, it requires skill, expertise, experience and an understanding of the approach adapted by potential operators/purchasers (BAHA 1993).

The BAHA suggested that the valuer should look at future earnings and cash flow over a period of ten years with an estimate of reversionary value in the tenth year. Readers are referred to the BAHA publication: *Recommended Practice for the Valuation of Hotels*, for a more detailed discussion of the DCF approach to Hotel valuations and to the considerations of 'net free cash flows' and 'maintainable earnings'.

Valuers in this market segment must be experienced and will often specialise in a very narrow sub-market as detailed market knowledge of the players and the 'business activity' are such critical elements.

(The authors are grateful to M Green for permission to use extracts from his Sheffield Hallam University course material in this chapter.)

Summary

The profits method

This method and quasi-profits methods are the preferred methods for the valuation of property bought and sold as operational entities such as hotels, public houses, petrol filling stations and most real estate used for the leisure industry.

Each segment of the leisure industry is so specialist that only a few valuers have the knowledge and experience to act in the marketplace.

The method relies on:

- specialist business knowledge and the ability to read accounts
- an ability to assess sustainable profits and /or net benefits as a cash flow over a five to ten year period
- a thorough knowledge of the major operators in the specialist sector
- an ability to convert sustainable profit or net cash flow into an expression of capital value using capitalisation techniques or discounted cash flow
- broad comparators, such as price per bedroom, which can be used to gauge probable values in an active market should not be used as the sole method of valuation.

Part 3

Investment Property

Investment Analysis

13

The other side of the valuation coin is the analysis of property investment opportunities. It may be useful or even essential for the valuer to see an investment from this perspective in order to inform the valuation. This requires careful consideration of the relationship between the known and expected returns and costs. Expected returns can only be estimated by a consideration of the future. The future is unknown. This necessarily introduces the valuer to the concept of risk.

Risk and valuation

A property investment is an exchange of capital today (current purchasing power) for future benefits. These future benefits may be in the form of income or capital growth, or a combination of both. As indicated earlier in the text, this requires the valuer to have some regard for the future and, as has so frequently been said before, the only certain thing about the future is its lack of certainty. When an investor purchases future rights he has accepted 'risk'.

When a valuer describes a property investment as being 'risky', he/she is implying some relative measure of uncertainty about the expected returns.

- The rents expected in the future may not be realised, so that for example expected rental growth will be less than anticipated.
- Increases in rent will not occur at the time expected or the property may become vacant and take some time to re-let.
- The capital value of the property on re-sale may not be realisable, may not increase with time or may fall with time.
- Costs associated with holding the property, such as repairs, may be unexpectedly high.

As previously noted these property risks may be systematic or unsystematic. Tenant risk, sector risk, planning risk and legal risk are unsystematic and, in a portfolio context, risk reduction can be achieved through diversification. Taxation, legislation, and structural risks are more systematic and cannot be actively reduced through diversification. But at an individual property level the investor is subject to specific or unsystematic risk, and in valuation the valuer is seeking to reflect the market's view of the risk relating to a specific property.

The more detailed the valuer's research into the financial stability of current tenants, the physical structure of the building, regional economics, and so on, the more able the valuer will be to express an expert opinion of OMV. The less certain the market is in respect of all these factors then the more risky the property will appear to be.

For this reason a valuer must reflect this greater risk in the valuation. The conventional income capitalisation approach adjusts for this through using a higher all-risks yield based on market evidence of sales of higher risk properties (see Chapter 5). In growth-explicit discounted cash flow approaches a common approach to dealing with risk is to increase the discount rate.

The use of a Risk Adjusted Discount Rate (RADR) is still the most popular market approach used in growth-explicit discounted cash flow approaches. However, this leaves the valuer largely reliant upon a subjective or intuitive adjustment based on experience or market knowledge. How much should the discount rate be increased to allow for risk? Such an adjustment can be arbitrary. Unless market participants are familiar and comfortable with explicit valuation approaches, and exchange assumptions about rental growth rates and discount factors, the discount rate is a subjective input.

The alternative is to use statistical techniques to examine the investment risk, more specifically to model cash flow uncertainty. One such approach which is sometimes adopted is scenario analysis, whereby the valuer makes three estimates of the future — the best estimate, the most likely, and the worst — and estimates the probability of each occurring. The valuation would be an average of some sort, but adjusted by the range of the alternatives (a wide range suggests a risky investment and a lower valuation).

However, this approach ignores all the other possible outcomes. Consider the following example (taken from Byrne and Mackmin, 1975).

You are instructed by a banking organisation to prepare a valuation for mortgage purposes of a new owner-occupied office building.

Having measured and surveyed the building and checked your findings with the architect's plans, you are satisfied that the building contains a total lettable area of 12,000m². It is your considered opinion, having regard to all the relevant factors, that the property would let at a figure of between £250 and £300 per m² and would sell as an investment on the basis of a 5–6.5% capitalisation rate.

A preliminary valuation is prepared as follows:

Area:	12,000 m²
Full rental value:	£3,100,000 (approx £260 per m²)
Yield:	6%
Income	£3,100,000
YP perp @ 6%	16.67
Valuation	£51,677,000

But this income capitalisation provided very little information. What is the expected cash flow? How risky is it? An explicit discounted cash flow valuation should involve some consideration of the variability of cash flows which is possible.

The ranges suggested for rents in the example is £250 to £300 per m² and the yield range is 5–6.5%. Given these limits, and taking steps of, say, £5 in rent and 0.25% in yield, Table 13.1 shows the variations in final valuation obtained by altering these two variables within their respective ranges.

Table 13.1 (values in £m)

Rental (£/m²)	Yield (%)						
	5.0	5.25	5.50	5.75	6.0	6.25	6.50
250	60.00	57.14	54.55	52.17	50.00	48.00	46.15
255	61.20	58.29	55.64	53.21	51.00	48.96	47.08
260	62.40	59.43	56.73	54.26	52.00	49.92	48.00
265	63.60	60.57	57.82	55.30	53.00	59.88	48.92
270	64.80	61.71	58.91	56.44	54.00	51.84	49.85
275	66.00	62.86	60.00	57.39	55.00	52.80	50.77
280	67.20	64.00	61.09	58.43	56.00	53.76	51.69
285	68.40	65.14	62.18	59.48	57.00	54.72	52.62
290	69.60	66.29	63.27	60.52	58.00	55.68	53.54
295	70.80	67.43	64.36	61.57	59.00	56.04	54.46
300	72.00	68.57	65.45	62.61	60.00	57.60	55.38

There are 77 possible 'outcomes' in Table 13.1. Which one is correct? Are any of them correct? Can the valuer justify his best assessment of £51.67m — which is clearly only one of a much larger number of possible solutions — when this selection also implies the conscious rejection of, in this case, at least 76 other values? Is it possible to use the information at our disposal to arrive at a closer estimate of the likely value of this property?

Initially, a range of values of between £46.15m and £72.00m may be noted. This range has a mean and a standard deviation. The latter is an accepted measure of risk. Can we use the range to define and adjust for the risk of the investment?

If an analysis of the data is made it may be possible to determine the relative frequency of occurrence for the various rental levels between the minimum of £250 and maximum of £300.

Let us suppose that for this example such an analysis is possible for 50 comparable transactions: the results can then be tabulated, as in Table 13.2. This gives a good indication of the probability of occurrence of the various possible rentals presupposing, of course, that the 50 transactions are truly comparable. If insufficient data are available it may be necessary to use other methods, as described below.

Each rental may now be 'weighted' by multiplying it by its probability of occurrence and the results summated to give one overall expected rental value, each element being included in proportion to the probability of its occurrence. All possible rental values will then have been built into the result. None is discarded at this stage, but their importance is now related to the known frequency of occurrence of each rental level. Since the distribution of probabilities in Table 13.2 is almost symmetric, then the expected rental will be in the centre of the distribution. In this case it is £275.

The expected rental obtained here is specific to the distribution shown in Table 13.2; other shapes of probability distribution can occur, and when this happens the expected value will be different. An analysis of transactions, for example, might show a different frequency and probability pattern, as in Table 13.3.

(It is not unreasonable to argue that in each of these distributions the 'modal value' — that rental having the largest observed frequency of occurrence — could be taken as representative, since it is the most likely value.)

Table 13.2

Rental (£/m²)	Frequency	% Occurrence	Probability
250	1	2.0	0.02
255	2	4.0	0.04
260	5	10.0	0.10
265	6	12.0	0.12
270	7	14.0	0.14
275	9	18.0	0.18
280	7	14.0	0.14
285	6	12.0	0.12
290	5	10.0	1.10
295	1	2.0	0.02
300	1	2.0	0.02
Total	50	100.0	1.00

Table 13.3

Rental (£/m²)	Frequency	% Occurrence	Probability
250	2	4.0	0.04
255	5	10.0	0.10
260	20	40.0	0.40
265	10	20.0	0.20
270	4	8.0	0.08
275	2	4.0	0.04
280	3	6.0	0.06
285	2	4.0	0.04
290	1	2.0	0.02
295	1	2.0	0.02
300	0	0.0	0.00
Total	50	100.0	1.00

In Table 13.3, the relative frequency of rents shows that some rentals are quite probable, £265 for example. The use of the expected value — the weighted mean — reflects the possibility that these other results might occur. Whereas the central value appears to be £275, the expected rental value in this case is £265, showing that, in spite of the evidence that 40% of observed rentals are at £260, 46% are above £2.60, and 14% below.

The different results from these two sample rents are compared in Table 13.4 for the seven possible yields used before. This range of values may be acceptable if the number of alternatives remains relatively small. The range of possible alternatives may be very large, however — and, more importantly, it may be possible to determine how likely they are to occur by means of an analysis of observed frequencies.

Table 13.4 (values in £m)

| Yield (%) | Rental (£/m²) | |
	£275	£265
5.00	66.00	63.60
5.25	62.86	60.57
5.50	60.00	57.82
5.75	57.39	55.30
6.00	55.00	53.00
6.25	52.80	50.88
6.50	50.77	48.92

Tables 13.2 and 13.3 show two different distributions of values. The first is close to a normal distribution; the second is clearly non-normal or skewed (negatively, or to the low side). Non-normal distributions are very common in property and make life more difficult, because the central value may not be the weighted average or expected value.

There are three possible ways of dealing with this problem:

* ignore it
* use a computer to calculate and display the results for all possible alternatives (such an exercise is called 'a simulation')
* make use of available experience to determine subjective probabilities for the occurrence of 'likely' values for these variables. These probabilities can then be built into the consideration of alternatives.

The final method is usually more likely to appeal to valuers and investors. We therefore develop this approach.

In the example described in Tables 13.2 and 13.3, the various yields were all suggested as equally possible. Clearly the valuer should be able to say from his knowledge that all are not equally possible, but that some are most unlikely and, more important, that some are very likely. This view is based upon the considered opinion of the valuer.

There are standard and easily-learned rules to enable such considered opinions to be converted to subjective probabilities recognised to be as valid as the objective assessments derived from long-run frequencies as shown in Tables 13.2 and 13.3. A complete probability distribution can be built up for any variable using these methods.

As has been seen earlier, yields are just as likely to vary as rentals in this example. After consideration, the following subjective probabilities have been placed against the possible yields (see Table 13.5).

The distribution of probabilities is such that only a few yields are considered likely to occur. From this distribution the expected value for the yield may be obtained by weighting each yield by its probability: in this case the value is 5.995% (6.00%).

The implication of the subjective selection of a high probability of occurrence for a particular yield is that it is considered relatively risk free. It is also possible, therefore, to use such assessments as indicators of individuals' attitudes to risk in their investment.

Table 13.5

Yield (%)	Probability
5.00	0.02
5.25	0.03
5.50	0.05
5.75	0.15
6.00	0.45
6.25	0.20
6.50	0.10
Total	1.00

The years' purchase may then be calculated, and the capital value arrived at in the usual way.

*Distribution I (*Table 13.2)
Estimated yield	6%	
Estimated full rental value		
12,000 × £275	£33,000	
YP perp @ 6%	16.67	
		£550,110

Distribution 2 (Table 13.3)
Estimated yield	6%	
Estimated full rental value		
12,000 × £265	£31,800	
YP perp @ 6%	16.67	
		£530,106

These valuations may be compared with the original best estimate or preliminary valuation.

Any solution derived in this way must be understood to be an estimate. The use of a statistical analysis of this type can only produce estimates. But — and this is more important — it produces a more consistent approach to uncertain situations, highlighting the stages in the appraisal process and pointing to any inconsistencies requiring correction or modification.

The method outlined is part of a more scientific approach to valuation (see Baum and Crosby, 1995, for a more detailed analysis of this type).

Every input — variable or otherwise — is dependent upon the strength of the valuer's evidence, his rental and value analysis, and his understanding and assessment of current market conditions. The approach differs in that the valuer is required to consider ranges of uncertain variables much more carefully, any single 'figure' arrived at being recognised as an estimate based on a proper analysis of the market.

Using a method such as this, a valuer may quite reasonably derive a series of results for any valuation. In that case, although there is no reason why a range of valuations may not be very helpful, great care must be taken in presenting such findings to the client who expects a single valuation result. The complete findings should be incorporated into valuation reports as appendices, as they are a full summary of the valuer's views of the investment.

Naturally the method should not be applied automatically. It requires a clear understanding of the statistical methodology and its implications; and it could also be inappropriate in some situations.

Many valuers see little need for the explicit use of probability in their valuations. In this book it has been indicated that on many occasions there is a lack of certainty. One cannot be certain what the rent of a vacant building really is until it is actually let, one cannot be certain what rent will be achieved on review, one cannot be certain of the capitalisation rates to be used, and one cannot be certain about future costs of repairs and refurbishment. There is a risk: why not reflect it?

In the market the use of simulations has been largely restricted to development appraisal work. Sophisticated approaches are now in place in many property organisations; statistical packages such as Crystal Ball and @RISK are available off the shelf for model-building; and property cash flow simulations are available through software such as OPRent.

None the less, valuers are naturally resistant to such approaches. An argument for adhering to market capitalisation approaches rests with the well-established view that accuracy in assessment of the key variables at today's date is sufficiently problematic without adding the difficulty of estimating future rental levels and capitalisation or discount rates. The further the valuer ventures away from the market, the greater is the probability that one is adding human error to the already high levels of risk and uncertainty that attach to the property.

None the less, valuers are regularly encouraged to embrace this type of technology. Baum and Crosby (1995) explore the use of sensitivity analysis. This they regard as 'a somewhat rudimentary risk analysis technique which helps investors to arrive at a decision but fails to identify the chances of the possible variations becoming fact'. Recognising the popularity of the RADR approach, they also explore and recommend the use of 'Certainty Equivalent Cash Flow Models', in particular the use of the hybrid 'Sliced Income Approach'.

These models make use of normal distribution theory which holds that some 68% of all values in a normal distribution will be within ± 1 standard deviation of the mean; 95% within ±2 and 99% within ±3 standard deviations. From this a certain equivalent cash flow can be constructed. Both Baum and Crosby and Dubben and Sayce (1991) apply this approach to property investment cash flows.

Given the best estimate of rent ±1 standard deviation, the best estimate of growth +1 standard deviation and the best estimate of yield ±1 standard deviation and a risk free discount rate, a Certainty Equivalent Cash Flow can be constructed.

The following is based on Dubben and Sayce's example of a property let at £10,000 for two years with an ERV of £15,000. Given this data, the cash flow could be considered as:

Current rent: £10,000 pa for 2 years

ERV: £15,000 pa minus 1 standard deviation (£1,000)
CE = £14,000

Rental growth: 5% minus 1 standard deviation (1%)
CE = 4%

Capitalisation rate: 6% plus 1 standard deviation (0.5%)
CE = 6.5%

It will be noted that this is a risk-averse method, as negative scenarios have been adopted for each variable. This, it is suggested, provides a Certainty Equivalent Value.

There are doubts about the usefulness of this technique when weighed against market experience. What (for example) is the certainty equivalent of a fixed contracted cash flow paid by a FTSE 250 tenant? The Sliced Income Approach is a more logical model for the typical property investment as it accepts the fact that the current rent payable under a lease is relatively risk free. The riskiness is likely to be a function of the tenant's ability to pay the rent, which can be judged reasonably accurately given a thorough assessment of the tenant's credit rating. This method allows the income to be split between the current rent which is certain, and the future rent which is less certain. Hence, given the underlying criteria of the model it is possible to assess the value of the certain rent at the risk free rate.

There is little doubt that if valuers use discounted cash flow, or growth-explicit, approaches to valuation, RADR approaches are clearly preferred in practice — but this is not to say that risk adjusted cash flows are not helpful in property investment analysis, and more examples of applications will emerge in time.

Risk and uncertainty in analysis

The phenomenal growth in institutional investment in land and buildings in the 1970s turned the attention of valuers to investment analysis. Nonetheless, the violent boom and slump of property prices in the early 1970s, the late 1980s and again in the central London office markets of 2000–2003 suggested that the market still had much to learn about the relationship between economic activity and the property market.

When property is purchased as an investment it is purchased for its present and future income and for capital growth. Occasionally other factors (such as prestige) enter into the decision, but for the rational investor such factors should not override decisions based on sound appraisal.

The following information may be required by investors, and is usually presented in a written report (see Appendix B).

Location

Here the valuer provides a description of the town, its geographical location in relation to other major towns and cities, details the population and road, rail and air communications, and comments upon the town's economic base. The position of the property within the town or city is then outlined.

Description of the property

This section of the report deals with the age, design, construction and condition of the building(s) comprising the subject property. A comment may also be made concerning running costs.

Town planning

A full report of the result of enquiries made with the local planning authority should be included.

Highways

Existing and any proposed changes or improvements to highways and the potential impact on the property must be noted.

Rating assessment

Rateable value (RV) and business rates payable are generally specified.

Tenure

Freehold or leasehold, full details of relevant covenants in leases, unusual or restrictive clauses in any ground rent review mechanism which could impact upon value and full details of all freehold title restrictions must be noted.

Details of occupational tenancies

Leases and lease terms in the case of freeholds must be specified as they may directly affect the market value. Information should include comments on the tenants' covenant following investigation of the tenant's financial standing.

Fire insurance

The extent of coverage required and premiums payable should be given together with details of premium recovery from tenants where applicable.

Management problems

All property requires management. In many cases the main costs, as in the case of a retail centre, will be recovered through a service charge. While there is likely to be a substantial non-recoverable cost this will need to be assessed, as it will reduce the net income.

Terms of the transaction, including the valuer's recommendations

Market analysis and context

While much of a valuer's report is descriptive and capable of factual accuracy, investors now require a much fuller statement to be made by the valuer as to the current and probable immediate future state of the market.

This will include some supportive data on current activity, yields and market rentals and is a clear necessity when a valuation for purchase and investment advice is sought.

This part of the report must include details of the current income and future rental growth prospects, the latter being substantiated with economic projections of the town and region's growth. Future rental growth will in part be influenced by the quality and/or suitability of the building for its permitted use.

In addition to the foregoing, investors require information on the rate of return or yield that they can expect from the investment. As indicated in Chapter 4, DCF techniques have been developed for analysing investments.

Where a valuer is asked to advise on a specific property investment for a specific client, the prospective purchase price is generally a known factor, and as a first step in an appraisal all acquisition costs such as solicitors' fees, survey and valuation fees and stamp duty, must be added to the estimated purchase price to assess the total outlay.

Yields

Some reference has already been made to problems of terminology relating to yields (see Chapter 5). As has been already suggested, there are many confusions concerning return measurement in property. This is largely due to the unique terminology which has developed in the property world; it is also due to the unique nature of property, especially its rent review pattern and the resulting reversionary or over-rented nature of interests.

There is also some misunderstanding of the difference between return measures which are used to cover different points or periods in time. Return measures may describe the future; they may describe the present; or they may describe the past.

Measures describing the future are always expectations. They will cover periods of time and may, if that period begins immediately, be called ex ante measures. An example is the expected internal rate of return from a property development project beginning shortly; another example is the required return on that project.

Measures describing the present do not cover a period, but describe relationships existing at a single point in time (now). An example is the initial yield on a property investment; this is simply the current relationship between the rental income and the capital value or price.

Finally, measures of return describing the past, or ex post measures, are measures of (historic) performance. An example is the delivered return on a project.

The following definitions describe the future.

IRR or expected return

This is the expected return using estimated changes in rental value and yield. There are similarities in other investment markets: these include gross redemption yield, holding period return, IRR. The term 'equated yield' is sometimes used in property circles but serves only to confuse and should not be used.

Required return

This is the return that needs to be produced by the investment to compensate the investor for the risks involved in holding the investment. It is also called the target rate or the hurdle rate of return. It is usually assessed as the sum of a risk free rate, such as the redemption yield on government fixed interest bonds or gilts, and a risk premium (see Baum, 2002). In this book we also used the term risk-adjusted discount rate or RADR for this concept.

The following definitions describe the present.

Initial yield

This is defined as net rental income divided by the current value or purchase price. There are similarities in other investment markets: these include interest yield, running yield, income yield, flat yield, dividend yield.

Yield on reversion

This is defined as current net rental value divided by the current value or purchase price. There is no equivalent in other markets.

Equivalent yield

This is the average of the initial yield and the yield on reversion. It can be defined as the IRR that would be delivered assuming no change in rental value, but this has created difficulties in the case of over-rented properties. As in the case of all IRRs, the solution is found by trial and error. There is no equivalent in other markets.

Reversionary potential

This is the net rental income divided by the current net rental value or vice versa. There is no equivalent in other markets.

The following definitions are performance measures. They describe the past.

Income return

This is the net rent received over the measurement period divided by the value at beginning of the period.

$$IR = Y_{0-1}/CV_0$$

Capital return

This is the change in value over the measurement period divided by the value at beginning of the period.

$$CR = [CV_1 - CV_0]/CV_0$$

Total return

This is the sum of income return and capital return.

$$TR = [Y_{0-1} + CV_1 - CV_0]/CV_0$$

Time-weighted return

This is the single rate of compound interest which will produce the same accumulated value as would be produced by a set of different periodic interest rates.

The TWRR ignores the timing of cash injections and extraction. It is appropriate for quoted unitised and other commingled funds. It is an inappropriate measure where the manager has discretion over cashflows or cash investments.

Money-weighted return

This is an approximation to the IRR. It is appropriate for funds where the manager has control over cashflows. It takes account of the amount invested in each period (hence the name). However, its use is now wholly unnecessary as computing facilities now exist to make IRR calculations easy enough to avoid the use of this approximation.

Internal rate of return

This is the most accurate and complete description of historic return. It is appropriate for managers who have discretion over the cashflow and takes into account the cash invested in each period. It is not, therefore, a mean of annual returns.

Given the total acquisition cost, the most simple level of analysis is to assess the initial yield. As defined above, this is the simple relationship or ratio of the first year's income and the total acquisition costs.

$$\frac{\text{Income} * 100}{\text{Acquisition cost}} = \text{Initial yield}$$

This is known as or equivalent to the interest-only yield, the flat yield or the running yield in other investment classes.

To take an example, assume a total purchase outlay of £1.25m and an initial rent of £100,000. The initial yield is £100,000 * 100 / £1.25m = 8%.

But 8% is not the expected return or IRR. Simplifying, the expected return will be higher than the initial yield by the amount of annual rental growth expected in the long term. Where the initial yield is i, the IRR is r and growth is g:

$$r = (1 + i)(1 + g) - 1$$

Assuming expected rental growth of 10%:

$$r = (1.08)(1.1) - 1 = 0.188 = 18.8\%$$

The expected return is then 18.8%.

Property let at its rack rental with rent reviews every five years restricts the potential receipt of any increase in rent to the rent review date. Although rents may rise continuously at some long run average compound rate, for property investments to equate with yields from stocks and shares the growth between reviews must be sufficient to compensate for the contracted fixed income between reviews.

Hence the purchase of a property let at £100 pa for five years with reviews assumed to be every five years for the sum of £1,250 must, if the discount rate is 18.8%, reflect a rate of growth in income greater than 10% pa compound. What rent is required at the next rent review?

The problem can be phrased as:

$$£1,250 = £100 \times \frac{(1 - PV)}{r} + \left[x \times \frac{1}{i} \times \frac{1}{(1 + r)^n} \right]$$

where $r = 18.8\%$ and $i = 8\%$ and x = rent on review.

$$1{,}250 = 100 \times \frac{1 - \dfrac{1}{(1 + 0.188)^5}}{0.188} + \left[x \times \frac{1}{0.08} \times \frac{1}{(1 + 0.188)^5} \right]$$

$$1{,}250 - \left[100 \times \frac{1 - \dfrac{1}{(1 + 0.188)^5}}{0.188} \right] + \left[x \times \frac{1}{0.08} \times \frac{1}{(1 + 0.188)^5} \right]$$

$$1{,}250 - [100 \times 3.074] = x \times 12.5 \times \frac{1}{2.366}$$

$1{,}250 - 307.40 = x \times 12.5 \times 0.422$

$942.60 \quad = \quad 5.27x$

$£178.59 \; = \; x$

Thus the rent on review in five years' time must be (say) £179 in order that the investment can be considered to be as acceptable as a continuous growth income of £100 pa compounding at 10% pa.

$£100 \times$ amount of £1 $\quad = \quad$ £179

Amount of £1 in 5 years $\quad = \quad \dfrac{£179}{100}$

Amount of £1 in 5 years $\quad = \quad 1.79$

Rate of growth $\quad = \quad$ approx 12.5% $((1.125)^5 = 1.7821)$

Clearly, given any initial yield, required return and rent review pattern, the required growth rate can be calculated.

Most investors will require to know the initial yield on total acquisition costs, the reversionary yield (yield on reversion), the equivalent yield and the internal rate of return on specific assumptions.

All of these calculations can be carried out on a before or after tax basis, but from the point of view of a specific client or investor it is suggested that if the client is subject to tax of any form on income or capital gains from property then net of tax returns should be provided.

In Chapter 3 the valuation of a property let at £100,000 pa for two years with a current market rental value of £200,000 pa, was considered in some depth. The relationship between the all risks yield (or market capitalisation rate) of 5% and an expected or target rate of 12% can be analysed to assess the implied rate of rental growth. On a five year rent review pattern the implied annual rental growth is 7.6355%. This produced an implied rent in two years of £231,708 and a capital value on a modified DCF basis of £3,863,316. These figures would be rounded in the market but are used here to indicate the kind of information that an investor might require.

Acquisition costs

The majority of investors require yield calculations to be based on total acquisition cost. Solicitor's fees, surveyor's fees and stamp duty are the usual costs that need to be added to purchase price. In special cases it may be necessary to take account of immediate capital costs such as repairs, or, in the case of a vacant building, refurbishment and letting expenses. For example:

Purchase price		£3,863,316
Add for legal fees ⎫	say 5.75%	
Surveyor's fees ⎬	inclusive	
Stamp duty ⎭	of VAT	£222,140
Total acquisition cost		£4,085,456

Initial yield

The initial yield may mean the relationship between the before-tax net rents receivable during the first 12 months of ownership and total acquisition cost; the return expected during the investor's current financial year; or the relationship between current contracted rents and total acquisition cost. In the absence of specific direction from the client it would be normal to assess the relationship between current contracted rent and total acquisition costs. For example:

Current rent	£100,000
Less landlord's expected	
non-recoverable management	
expenses inclusive of VAT	
Minimum fee, say	£5,000
	£95,000

Initial yield = (£95,000 / £4,085,456) × 100 = 2.325%

This yield can be crucial to some investors. Equity investors — those using their own capital to purchase, such as insurance funds and pension funds — may need a minimum income yield to buy a property asset. Debt investors — those borrowing a large proportion of the acquisition costs — will need a minimum income return to pay interest charges.

Other funds, while having regard to the initial yield and its implication for their whole portfolio, will be aware that a low initial property yield can be balanced by income growth to achieve a total return for the fund above the target rate.

Yield on reversion

Debt investors may be able to finance a shortfall in rent for a period of time, but will be interested to know when holding reversionary property whether, on conservative assumptions, they will be able to cover interest charges when the reversion is due and a full market rent is paid. It is therefore normal in an investment report to include an assessment of the yield at reversion calculated on the basis of current estimates of open market rentals. For example:

	Estimated rental value	£200,000
Less	Non-recoverable management fees say	£7,500
		£192,500

Yield on reversion = (£192,500 / £4,085,456) × 100 = 4.71%

Equivalent yield

Most investors will require advice on the internal rate of return based on current rentals and values. This is another conservative or risk-averse way of looking at the investment, with similarities to the certainty equivalent approach described above. It is more appropriately thought of as an average of initial and reversionary yields.

A common approach to estimation of the equivalent yield where the value is given is to use trial and error and a term and reversion valuation with a single, unknown yield throughout. The yield or capitalisation rate which produces the capital value of the investment given the current rent, the current rental value and the term to reversion is the equivalent yield (see example 6.10 p109).

Expected return

Investors will require an estimate of the expected return. This also helps the investment valuer in formulating his or her investment purchase advice. For example:

Implied rent in 2 years	£231,708
× YP perp at 5%	20
	£4,634,160

Here the IRR on an annual in arrears assumption is found from

0	−£4,085,456
1	+£95,000
2	+£4,729,160 (£4,634,160 + £95,000)

and is 8.758%

This variation highlights some very important but as yet unresolved issues in the professional areas of valuation and investment advice. It has historically been assumed that the 5% all risks yield would account for the non-recoverable management costs, and no account was taken of acquisition costs.

The profession is still divided over the correct approach in respect of valuation work. Investment advice will normally reflect net income as seen by the prospective purchaser and total acquisition cost as likely to be incurred by the purchaser.

A strongly held view is that:

(a) management costs should always be deducted in all valuations and
(b) acquisition costs should be deducted from the valuation figure before expressing an opinion on value.

Note the effect this has in the following amended valuation and analysis.

eg   Revised valuation:

Next 2 years	£100,000	
Less management, say	£5,000	
	£95,000	
YP 2 years at 12%	1.6901	
		£160,559
Reversion to	£231,708	
Less management, say	£7,500	
	£224,208	
YP perp @ 5%	20	
	£4,484,160	
PV £ at 12% for 2 years	0.79719	
		£3,574,727
		£3,735,286

Let x = purchase price and acquisition costs at 5.75% =   0.0575x

Then $x + 0.0575x$ =   £3,735,286

       1.0575x =   £3,735,286

       x =   £3,532,185 the corrected valuation figure

Revised expected return analysis:

0 – £3,735,286   (£3,532,185 + 5.75%)

1 + £95,000

2 + £4,579,160   (£224,208 × 20 YP + £95,000)

which gives an IRR of 12%

How much information the valuer provides the investor with will depend upon the investor's requirements. The assessment of implied rental growth rates allows the valuer to compare historic growth patterns in the specific market with current market expectations and with the growth needed to achieve an overall acceptable return for the investor.

Investment worth

A number of writers believe that investment advice needs to be based on a more rigorous cash flow approach. This process can be very simple or very sophisticated, but in either form provides the analyst with the opportunity to test the sensitivity of the investment to a variety of variables, including rent review costs and lease renewal costs; voids and refurbishment costs; other non-recoverable service or repair costs; sale costs; income and capital gains tax; depreciation; exit yields and residual values. Such modelling also allows the valuer to test the effect of different rental growth factors either on an average long-term basis or by adopting different rates of growth over different time periods.

RICS Books has published an information paper on the Calculation of Worth. A worth calculation in effect looks at an investment from the owner's or potential owner's perspective. A cash flow as described above is prepared over a holding period which is typically taken as ten years but will be specified by or agreed with the client. The cash flow incorporates as many or as few of the above variables as may be required by the client but in a worth calculation a critical input will be the client's assessment of growth in market rents or a simulation agreed with the client to examine 'what if' market rents grow by 1% through to 5%. A simple spreadsheet structure for a worth calculation will be found at the end of the chapter.

A worth calculation is carried out to support an investor's decision to buy or an owner's decison to sell. In effect this is how the client with their property investment advisors determines whether a property being offered at a given asking price is in financial terms worth purchasing, or, given an annual review of a portfolio's market values, whether there are properties which could be sold at figures above the portfolio managers own assessment of worth.

It has already been indicated that the complexities of certain investments can only effectively be handled in this way. The complexities of current development funding schemes and the risks involved also suggest that the valuation and analysis of these kinds of property investments will also have to be handled by a form of simulation.

The importance of risk means that the probability of achieving the stated yields or returns will need to be estimated. Consider Example 13.1.

Example 13.1

Your clients are contemplating the acquisition of a property investment. The property is fully-let, producing a net cash flow from contracted rents of £1,000 pa for the next two years. These rents are considered to be certain. In two years' time the rents are due for review. You have considered the likely level of rents in two years' time and expect the rent roll to be in the region of £3,000; expressing this in capital terms you have prepared the following table of cash flows together with a probability measure for the capital reversions in two years' time. The asking price plus acquisition cost gives a total purchase price of £32,640. Advise your client.

End year	Net sum	Probability	Expectation	
1	£1,000	1	£1,000	
2	£1,000	1	£1,000	
	£30,000	0.20	£6,000	£36,500
	£37,500	0.60	£22,500	(Expected return
	£40,000	0.20	£8,000	in two years' time)

To assess the internal rate of return the generalised formula

$$\frac{I_1}{(1+i)} + \frac{I_2}{(1+i)^2} + \ldots \frac{I_n}{(1+i)^n}$$

must be solved for i, in this case in terms of:

$$£32,640 = \frac{1,000}{(1+i)} + \frac{1,000 + 36,500}{(1+i)^n}$$

The internal rate of return is virtually 8%. If the client's target rate is 7% the NPV can be calculated as follows:

Year	Sum	PV at 7%	
1	£1,000	0.9346	£934.60
2	£37,000	0.8734	£32,752.50
			£33,687.10
	Less purchase costs		£32,640.00
	Positive NPV say		£1,047.00

The investment on this basis is worthwhile.

Such calculations should of course be carried out on a true net basis from the specific client's viewpoint.

Here it has been recognised that the position in two years' time is not certain, as there is a rent review. In this example the rent on review has been assumed to be certain at £3,000, but the future benefit of £3,000 in perpetuity may have a present worth of between £30,000 and £40,000 depending upon the analyst's assumptions as to the discount rate. The 'expected value' in two years' time has therefore been calculated before considering the internal rate of return on this basis. The normal analysis would have opted for the most probable figure of £37,500 and would have ignored the other possible outcomes. A certainty equivalent approach could be used.

It is essential for the prudent investor to pause and consider the other possible outcomes and the effect that a change in the reversionary value might have on expected return. In this example it has been stated that the rent will be £3,000, but the market capitalisation rate of 8% could, under certain circumstances, fall as low as 7.5% and, under others, rise as high as 10%.

These forecasts would be based on a reasonable interpretation of the state of the economy and the extent to which national and local economic factors might affect future capitalisation rates for this type of investment. It will therefore be desirable in some instances to consider not only the possible rates of return but also their probability, and hence the variance of the return and the standard deviation.

Table 13.6

Possible rate (%)	Probability of occurrence	Expected rate of return %	Deviation of possible from expected	Deviation squared	Deviation squared × probability
7.5	0.2	1.5	−0.8	0.64	0.128
8.0	0.6	4.8	−0.3	0.09	0.054
10.0	0.2	2.0	1.7	2.89	0.578
	1.0	8.3			0.760%

Table 13.6 indicates that the expected rate of return is 8.3%, this being the weighted average of all possible outcomes which, in relation to an income of £3,000, would give an expected value of £36,144

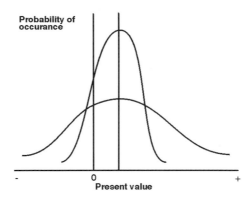

Figure 13.1

(YP × income). The variance of return in this case is 0.76%, which can be converted into a standard deviation by taking the square root, giving a standard deviation of 0.87%.

These latter two measures are measures of the risk involved. The smaller the standard deviation the smaller is the probability of the actual return deviating from the expected rate of return. In certain cases such a measure is extremely useful, particularly if one has an extreme case of two properties both with an expected rate of 8.3%, but one with a standard deviation of say 0.87% and the other with a standard deviation of 3%, for it is clear, in these terms, that the former is a far less risky investment.

An alternative is to consider the probability distribution of present values for alternative investments. For example, given distributions as in Figure 13.1 the less speculative investment is investment Λ. The client is therefore in a position to make a better decision between investments than he would have been if the only information he had received was the figure of present value. He can also judge whether the added return from the more speculative investment is likely to compensate for the added risk.

Probability distributions as in Figure 13.1 help to emphasise the possibility of a negative present value. Different investors have different attitudes towards risk and some cannot afford to make an investment if there is the slightest possibility of a loss. So to advise on the purchase of a property merely because it shows a positive NPV at the investor's target rate of discount is inadequate advice, because it again ignores the other possible outcomes which could occur with a change in income forecasts. Some indication of this possibility should be given to the investor.

A further alternative for considering uncertainties in investment decision-taking is the decision tree. Here again informed guesses as to probable outcomes need to be made.

In Example 13.1, while a number of probability factors were incorporated, these were based on a specific assumption as to a future event at the end of year 2, namely that the income would be £3,000 per annum. In reality this will be uncertain. Thus at the end of year 2 the tenant could renew the lease or he could vacate the property, and in the latter case it might either be re-let immediately, or there might be some probability that it would take a year to find a new tenant.

Provided probability factors can be assessed for each of the alternative outcomes a decision tree can be constructed. With this information the valuer is able to assess the present value of the entire tree.

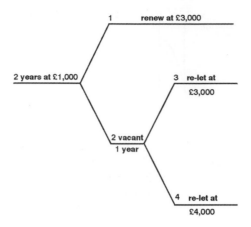

Figure 13.2

Present valuation calculations on an 8% basis

Branch 1

Fixed income for 2 years	£1,000	
PV £1 pa for 2 years at 9%	1.7833	£1,783
Re-let at	£3,000	
PV £1 pa in perp at 8%X		
PV £1 in 2 years at 8%	10.71674	£32,150
		£33,933

Branch 3

Income of £1,000 for 2 years as before		£1,783
Re-let at	£3,000	
PV £1 pa in perp at 8%X		
PV £1 in 3 years at 8%	9.9229	£29,769
		£31,552

Branch 4

Income of £1,000 for 2 years as before		£1,783
Re-let at	£4,000	
PV £1 pa in perp at 8%X		
PV £1 in 3 years at 8%	9.9229	£39,691
		£41,474

This investment has a present value at 8% of £33,933, or £31,552, or £41,474. Allowing for the probabilities and conditional probabilities we have:

£33,933 × 0.08	=	£27,146
£31,552 × 0.2 × 0.6	=	£3,786
£41,474 × 0.2 × 0.4	=	£3,317
Weighted present value:		£34,249

This can be compared to the asking price of £32,640. These techniques have long been used in business management but are still in their infancy in the valuation profession. However, packages such as OPRent now mean that there is commercially available valuation software which uses this approach.

The problems of how to fully reflect risk in analysis increases with the complexity of the problem itself. The simplest decision is in respect of the single investment opportunity and the yes/no decision. From that one moves to the decisions which involve choices between alternative investments, after which one becomes inevitably involved in portfolio risk analysis. An investor must concern himself not only with the risks of a proposed new investment, but also he must consider the impact of that new investment on any existing investments held within a portfolio.

Thus a valuer advising on a property might view it as a high risk and advise against the purchase, but the addition of that investment to an existing portfolio might make it a perfectly acceptable risk. The subject of portfolio analysis is outside the scope of this book.

Future uncertainty continues to be the reason given by most valuers for restricting the amount of detail provided for investors to a minimum. However, a number of investors currently expect their investment surveyors to provide them with more information on yields and more supportive data on market rents, market analysis and market performance. This information is used by the investors in their own assessment of the probability of achieving their investment aims.

Similarly, banks are now expecting valuers to provide much more detail. For example to set out explicitly their current opinion of the market. An open market valuation in accordance with the RICS definition is only a snapshot of the market and provides the client with an expression of the best price the property could have been expected to sell for at the valuation date, given proper prior marketing.

On its own, this can be likened to a photograph of a ball in mid air. The problem is that (except with exceptional photography) the untrained eye cannot judge whether the ball is going up or coming down. The same is true of the market value snapshot where investors and lenders are now concerned to discover from the valuer a considered and supported opinion as to the state of the market in rental and capital terms.

Regression analysis

Regression analysis is a standard tool for financial analysis. In its simplest form, the statistical technique of regression analysis enables the analyst to predict the value of one variable from the known value of another. Valuation would be an extremely simple science if, for example, the valuer could predict the sale price of a property from its total floor area. This would indicate that a linear relationship between value and size existed which, if plotted on a graph, would produce a straight line, the slope of which would be the variable factor, b. Where a is a constant, the standard formula for a linear relationship is

$$y = a + bx$$

where, in the simple case selected above, y represents house value and x represents total floor area. For example, it might be noted that in a given locality the price of houses was always equal to £5,000 plus the floor area multiplied by 5. Then:

$y = £5,000 + 5x$

However, it is a normal presumption that value is a function of a number of variables. Multiple linear regression enables the analyst to bring into play as many variables that may be considered to be likely to affect the value or likely selling price of the subject property, such as parking facilities, outlook, location, size, specific facilities such as central heating and garage space and other factors such as the age and condition of the building.

Regression packages developed for property provide inputs consisting of a list of property characteristics for the type of property under analysis, together with details of the actual sale prices or rentals achieved, or repair costs incurred. The step-wise model correlates each feature with the known factor, sale price, selects those feature with the highest correlation, produces a regression equation and estimates the sale prices on that basis. The computer then calculates the difference between actual price and the estimated price, and then proceeds to select from the remaining features the next highest correlation and proceeds until all the features have been used.

The end result is an equation which can, with care, be used to estimate sale price, or whatever other factor is required in respect of another property. As more data becomes available these are added to the existing store and the program re-run to check for any significant changes in preferences by purchasers.

Any predicted figure must not be regarded as an absolute, and the valuer requires some indication of accuracy or acceptability.

The statistical measure is generally R (the correlation coefficient) and R^2 (the coefficient of determination).

When the data produces a value for R as close as possible to 1 this would imply that the variation of the dependent variable (sale price, rental value) is explained fully by the independent variable(s). If R falls below 0.9 then less than 80% of the variation of the dependent variable is explained by the independent variable(s) and the smaller R becomes the less meaningful is the whole analysis.

The use of multiple regression analysis has been primarily limited to mass valuation problems, particularly for land-tax purposes and residential property. The majority of reported examples concern house prices, but it is suggested that regression and multiple regression could be used for other applications, such as:

- testing the relationship between size and price or rent (for example, the rate at which land price per hectare decreases with the increase in the size of the holding being sold, or the extent to which rents per m² for office or industrial space decrease with the size of the letting)
- determining the rental value of all types of premises (as rent is a function of size, location, facilities, running costs, consumer income)
- predicting gross trading income from licensed premises, theatres, restaurants, etc, from the number of persons using the premises
- predicting petrol throughput for service stations based on traffic counts and other variables
- estimating repair expenses based on property maintenance records.

Clearly valuations and analyses of the type outlined in this chapter are only likely to be undertaken by the larger national valuation practices, government and institutional investors. For these reasons the description of the alternative techniques used has been kept to a minimum to give the average reader and student valuer a general idea of the developing techniques. Readers requiring a more detailed approach to the techniques are referred to the recommended reading at the end of the book.

We began by suggesting that valuers should keep their minds open to new techniques. A major development in appraisal techniques in the near future will be the use of statistics, and Norman Benedict aptly summarised our own feelings in an article published in the *Appraisal Journal* as long ago as October 1972:

> Statistical analysis can significantly broaden the role of the appraiser and substantially increase his effectiveness, providing him with the tools to attain greater sophistication and expertise in the areas of marketability, feasibility and investment analysis.
>
> Correspondingly, the appraiser who clings to yesterday's tools to meet tomorrow's challenges will become progressively less effective and less involved in his work, while the appraiser who seeks constantly to acquire new skills will develop both personally and professionally. In summation, then, statistics represents a golden opportunity for an appraiser to experience both personal and professional growth.

This statement is as apt today as it was in 1972. Its pertinence is obvious to those valuers developing or using statistical or econometric models and willing to tackle the critical area of property market forecasting. There are many opportunities for practitioners to develop the proper use of these techniques in responding to their clients' needs. After all, an opinion of value is easily challenged; it will only be defensible if it is supported by best practice analysis of good available evidence.

Questions

Define:

(a) IRR
(b) initial yield
(c) equivalent yield
(d) required return

A property has just been sold for £1 million freehold. It is currently producing a rent of £60,000 pa net. The full rental value is £110,00 pa. The lease will be renewed in 4 years' time.

Calculate:

(a) total acquisition costs with fees and stamp duties @ 4%
(b) current return based on purchase price plus acquisition costs
(c) expected return after the lease is renewed
(d) the capitalisation rate(s) used by the valuer (the equivalent yield)
(e) the internal rate of return assuming rental growth @ 5% pa and a re-sale in 4 years' time, assuming no change in the capitalisation rate
(f) the growth in rental value needed over the next 4 years:
   (i) in pounds
   (ii) as a rate of growth per annum in order to achieve a required return (IRR) of 12%.

Spreadsheet User 13

The Excel spreadsheet facility now available to most valuers provides a flexible tool for addressing typical repeat client tasks. One such example would be the use for the purpose of calculating worth to an investor given agreed inputs as to:

- holding period
- rental growth rate
- clients target rate
- factual information as per current leases, rents, outgoings etc
- variables such a non recoverable management costs, lease renewal costs, rent review costs
- capitalisation rates
- Estimated Market Value or probable sale price at the end of the holding period.

Simulation will quickly convince most readers that the key variable in a growth calculation is the assumption relating to the rate of rental growth. Where possible this will be supported by market analysis, calculations of implied rates of rental growth and conclusions derived from econometric models.

The illustration shown here is very simple and purely provided to spur readers on to developing their own more user friendly models. In practice such calculations would be undertaken on a precise calendar basis, most commonly on a quarterly basis.

Microsoft Excel - Worth 2.xls

File   Edit   View   Insert   Format   Tools   Data   Window   Help

Q36

INVESTMENT WORTH CALCULATIONS

DATA							
CURRENT RENT					£100,000		
MARKET RENTAL VALUE					£150,000		
RENTAL GROWTH RATE					3.50%		
ALL RISKS YIELD AT END OF HOLDING PERIOD					7%		
INVESTORS TARGET RATE					10%		
MANAGEMENT FEES					£5,000		
RENT REVIEW FEES					5%		
LEASE RENEWAL FEES					7%		
RENT/LEASE RENEWALDATE						2009	2014
MANAGEMENT FEES INFLATION INCREASE ADJUSTMENT					3%		
DATE OF ANALYSIS						2006	

ANNUAL IN ARREARS ASSUMPTIONS

Period	MRV + Growth	Rent	Management fee	RR/LRFees	Net Income	PV at H9	PV M * K
0	£ 150,000		£ 5,000				
1	£ 155,250	£ 100,000	£ 5,150		£ 94,850	0.9091	£ 86,227
2	£ 160,684	£ 100,000	£ 5,305		£ 94,696	0.8264	£ 78,261
3	£ 166,308	£ 100,000	£ 5,464		£ 94,536	0.7513	£ 71,027
4	£ 172,128	£ 166,308	£ 5,628	£ 8,315	£ 152,365	0.6830	£ 104,067
5	£ 178,153	£ 166,308	£ 5,796		£ 160,512	0.6209	£ 99,665
6	£ 184,388	£ 166,308	£ 5,970		£ 160,338	0.5645	£ 90,506
7	£ 190,842	£ 166,308	£ 6,149		£ 160,159	0.5132	£ 82,187
8	£ 197,521	£ 166,308	£ 6,334		£ 159,974	0.4665	£ 74,629
0	£ 204,435	£ 197,521	£ 6,524	£ 13,826	£ 177,171	0.4241	£ 75,138
10	£ 211,590	£ 197,521	£ 6,720		£ 190,801	0.3855	£ 73,562
Sale price					£ 2,985,709	0.3855	£ 1,151,151

INVESTMENT WORTH   **£ 1,986,420**

The sale price at the end of year 10 has to be based on a valuation at that point in time .
In this case this is £197,521 for 3 years plus £211,590 in perpetuity deferred 3 years all at 7%.

Sheet1 / Sheet2 / Sheet3 /

Ready

Appendix A

Definition of Market Value

This definition of Market Value is taken from the RICS Appraisal and Valuation Standards.

PS3.2 Market Value

Valuations based on Market Value (MV) shall adopt the definition, and the conceptual framework, settled by the International Valuation Standards Committee.

Definition

> The estimated amount for which a property should exchange on the date of valuation between a willing buyer and a willing seller in an arm's-length transaction after proper marketing wherein the parties had each acted knowledgeably, prudently and without compulsion.

Conceptual Framework, as published in International Valuation Standard 1

3.2 The term *property* is used because the focus of these Standards is the valuation of property. Because these Standards encompass financial reporting, the term *Asset* may be substituted for general application of the definition. Each element of the definition has its own conceptual framework.

3.2.1 'The estimated amount ...'

Refers to a price expressed in terms of money (normally in the local currency) payable for the property in an arm's-length market transaction. *Market Value* is measured as the most probable price reasonably obtainable in the market at the date of valuation in keeping with the *Market Value* definition. It is the best price reasonably obtainable by the seller and the most advantageous price reasonably obtainable by the buyer. This estimate specifically excludes an estimated price inflated or deflated by special terms or circumstances such as atypical financing, sale and leaseback arrangements, special considerations or concessions granted by anyone associated with the sale, or any element of Special Value.

3.2.2 '... a property should exchange ...'

Refers to the fact that the value of a property is an estimated amount rather than a predetermined or actual sale price. It is the price at which the market expects a transaction that meets all other elements of the Market Value definition should be completed on the date of valuation.

3.2.3 '... on the date of valuation ...'

Requires that the estimated *Market Value* is time-specific as of a given date. Because markets and market conditions may change, the estimated value may be incorrect or inappropriate at another time. The valuation amount will reflect the actual market state and circumstances as of the effective valuation date, not as of either a past or future date. The definition also assumes simultaneous exchange and completion of the contract for sale without any variation in price that might otherwise be made.

3.2.4 '... between a willing buyer ...'

Refers to one who is motivated, but not compelled to buy. This buyer is neither over-eager nor determined to buy at any price. This buyer is also one who purchases in accordance with the realities of the current market and with current market expectations, rather than on an imaginary or hypothetical market which cannot be demonstrated or anticipated to exist. The assumed buyer would not pay a higher price than the market requires. The present property owner is included among those who constitute 'the market'. A valuer must not make unrealistic assumptions about market conditions or assume a level of Market Value above that which is reasonably obtainable.

3.2.5 '... a willing seller ...'

Is neither an over-eager nor a forced seller prepared to sell at any price, nor one prepared to hold out for a price not considered reasonable in the current market. The willing seller is motivated to sell the property at market terms for the best price attainable in the (open) market after proper marketing, whatever that price may be. The factual circumstances of the actual property owner are not a part of this consideration because the 'willing seller' is a hypothetical owner.

3.2.6 '... in an arm's-length transaction ...'

Is one between parties who do not have a particular or special relationship (for example, parent and subsidiary companies or landlord and tenant) which may make the price level uncharacteristic of the market or inflated because of an element of *Special Value* (defined in IVSC Standard 2, para. 3.11). The *Market Value* transaction is presumed to be between unrelated parties each acting independently.

3.2.7 '... after proper marketing ...'

Means that the property would be exposed to the market in the most appropriate manner to effect its disposal at the best price reasonably obtainable in accordance with the *Market Value* definition. The length of exposure time may vary with market conditions, but must be sufficient to allow the property to be brought to the attention of an adequate number of potential purchases. The exposure period occurs prior to the valuation date.

3.2.8 '... wherein the parties had each acted knowledgeably, prudently ...'

Presumes that both the willing buyer and the willing seller are reasonably informed about the nature and characteristics of the property, its actual and potential uses and the state of the market as of the date of valuation. Each is further presumed to act for self-interest with that knowledge and prudently to seek the best price for their respective positions in the transaction. Prudence is assessed by referring to the state of the market at the date of valuation, not with benefit of hindsight at some later date. It is not necessarily imprudent for a seller to sell property in a market with falling prices at a price which is lower than previous market levels. In such cases, as is true for other purchase and sale situations in markets with changing prices, the prudent buyer or seller will act in accordance with the best market information available at the time.

3.2.9 '... and without compulsion.'

Establishes that each party is motivated to undertake the transaction, but neither is forced or unduly coerced to complete it.

3.3 Market Value

is understood as the value of a property estimated without regard to costs of sale or purchase, and without offset for any associated taxes.

Appendix B

Illustrative Investment Property Purchase Report

The report here was based originally on material provided by Gooch and Wagstaff, Chartered Surveyors for the third edition. It has been amended by the authors and is retained solely as an example for student use.

In today's market, valuation reports, undertaken for Red Book purposes, must comply fully with all relevant sections of the Red Book. These same requirements are a recommended basis for all valuation reports.

A full report of a single tenanted commercial property with all appendices prepared in compliance with the Red Book will be typically 50–80 pages in length. Readers will appreciate that the report here would, in a real instruction, contain substantially more detail.

The authors are grateful to DTZ Debenham Tie Leung Ltd for allowing them sight of a current Red Book report of which no part has been reproduced or incorporated in this illustrative report.

PS5.1 Minimum Content of Valuation Reports

The report must deal with all the matters agreed between the client and the member in the Terms of Engagement and include the following minimum information, except where the report is to be provided on a form supplied by the client:

(a) identification of the client
(b) the purpose of the valuation
(c) the subject of the valuation
(d) the interest to be valued
(e) the type of property and how it is used, or classified, by the client
(f) the basis, or bases, of the valuation

(g) the date of valuation

(h) the status of the member and disclosure of any previous involvement

(i) the currency to be adopted

(j) any assumptions, special assumptions, reservations, and special instructions or departures

(k) the extent of the member's investigations

(l) the nature and source of information to be relied on by the member

(m) any consent to, or restrictions on, publication

(n) any limits or exclusion of liability to parties other than the client

(o) confirmation that the valuation accords with these standards

(p) the opinions of value in figures and words

(q) signature and date of the report.

In meeting the above requirements a valuer might construct a report under the following headings.

Executive summary

Terms of Engagement including basis of valuation, assumptions, special assumptions, and statement of compliance, where relevant, with the Appraisal and Valuation Standards.

Address of property

Property inspection and information

- Location
- Description and construction
- Accommodation and floor areas
- Condition
- Site
- Services
- Planning
- Rating
- Tenure
- Occupation leases
- Market commentary
- Market rental value
- Market yields
- Reinstatement cost (if required).

Valuation Limitations

- Confidentiality and non publication
- Any other conditional limitation agreed in the terms of engagement.

Appendices

- Copy of terms of engagement or instructions
- Definition of market value etc

- Plans — location, site, floor plates, pedestrian flows, GOAD
- Photographs
- Details of market rental analysis and valuation calculations.

Illustrative property acquisition report

Pension funds and life funds act in a trustee capacity when they invest policyholders' moneys. It is customary for an investment purchase report to be commissioned following agreement on purchase price to protect the trustee status. These reports are very similar to full reports commissioned for loan and other purposes. Each in turn will reflect the agreed instructions and will be prepared, where applicable, in accordance with mandatory requirements and/or general guidance notes issued by the RICS. Additional caveats will be agreed at the time of instruction and included in the report if so required by the valuer's Professional Indemnity Insurance Policy.

The format of these reports is as follows and will be supported by photographs, location plans, building plans, extracts from 'Focus' and other data bases to support opinions in the report including market evidence and opinion on rents and market capitalisation rates. All measurements used in reports should be stated to be approximate, and are now required to be in metric, or both metric and imperial. Technical language should be avoided but if used will need to be explained.

1.0 Summary

Terms have been agreed on behalf of XYZ Ltd to purchase this freehold shop investment from ...

Purchase Price:	£525,000 (five hundred and twenty five thousand pounds)
Tenure:	Freehold
Approximate Net Internal Area of ground floor:	100m^2
User:	High-class retail shop
Letting:	Let to AAA and BBB trading as ZZZ: Bakers on assignment from XXXX(UK) Ltd 25 years from 00/00/00 to 00/00/00, FRI terms £24,000 a year.
Reviews:	00/00/00 and every 5 years upwards only.
Estimated Market Rental Value:	£30,000 a year
Estimated Yields:	4.32% net initial rising to 5.40% on reversion in 00/00/00. Net equivalent yield 5.27%.

We conclude that the price agreed reflects the current market and we recommend the acquisition to XYZ Ltd.

This summary should only be read in conjunction with the detailed comments which follow:

> 1.0 Reports may be written in letter style or report format. Each page should be numbered and should bear the name of the valuer's practice. Paragraph numbering may avoid the risk of sections becoming detached and not missed and helps with cross referencing.

Address of property

2.0 Instructions

We are instructed to prepare a Report and Valuation on the freehold shop investment located at XXXX on behalf of XYZ Ltd. It should be noted that while we have inspected the property and briefly describe its structure, we are not instructed to report on its structural condition and we understand that you have carried out your own structural survey and are satisfied with the result.

2.0 In accordance with RICS requirements, terms and conditions will have been agreed in writing prior to acceptance of the instruction. This section would refer back to the details set out in the Conditions of Engagement. Where a structural survey is part of the same instruction the valuer's report should draw on the findings of the Building Surveyor's report.

2.1 This report is subject to the Limitations set out in section 16.

3.0 Location

XXXX is a historic City with a population in the region of 70,000 persons which is substantially increased by visiting tourists and by students. The estimated shopping catchment is some 200,000 persons. The City is situated close to the junction of the M0 and the A00 thus providing good access to London (100 km), XXX (12.5 km) and XXXX (6.25 km). The A00 to Z, the A 0 to B and C and the A000 to D all radiate from the Town thus providing good communications to the surrounding centres and to the motorway network.

A main line rail service connects XXXX to London Kings Cross (the fastest journey time being 2 hrs and 55 mins) and X Airport is approximately 25 miles to the North by road. A location map is attached as Appendix 2.

The property is located within the established shopping area in the City Centre on the South side of XXXX at the top of XXXX Street and is close to the Town's main car parks. A Street Map is attached as Appendix 3. The City Centre is well represented in terms of national multiples and includes such occupiers as Next, Boots, Marks & Spencer, FW Woolworth, Burtons and WH Smith. A market is held in XXXX Place on Saturdays and this is a busy pedestrian thoroughfare linking the Town's main car park with the principal shopping area of Market Place and High Street. An extract from the trader's plan is attached as Appendix 4.

3.0 An opinion as to the suitability of the location for the current use of the building will be given. In the case of retailing details, of pedestrian flows, travel time catchment area, spending power and competitive retail centres might be included. This is data to support subsequent opinions on the quality of the centre.

Alternative emphasis may be necessary for distribution, manufacture etc.

Employment indicators may be included.

4.0 Site

The site which slopes up to the rear is roughly rectangular in shape and has a frontage to XXXX of 4.35 metres and a depth of 31.6 m. The site extends to an area of approximately 126 m² and is outlined in red on the attached Ordnance Survey Extract (Appendix 5).

> 4.0 A plan will indicate the extent to which the site is fully developed. If the site is under-developed it should be mentioned here.

5.0 Description and construction

The property is a Grade II Listed Building constructed circa 1850 comprising a ground floor shop unit with rear storage and 3 floors of disused residential space above having access from the rear yard. Access to the yard is provided via a passageway to the side of the shop. The shop has a somewhat limited internal width averaging 3.2 m.

The building is constructed of brick and sandstone quoin blocks and eaves cornice under a pitched slate roof. The windows are single glazed wooden sash. The shop has no central heating. The upper floors are heated by a coke burning boiler in need of replacement, serving radiators.

The separate access to the upper floors means that the three floors of disused space could be brought back into use subject to necessary consents.

> 5.0 This section needs to be as factual as possible and should avoid personal judgements. The valuer must ensure that the coverage is not such as to suggest a structural survey has been undertaken, unless that is the case.
>
> Comments on eaves, heights, load bearing and loading facilities would be relevant for certain types of property.

6.0 Accommodation

We have measured the property in accordance with the RICS Code of Measuring Practice and the building has the following approximate floor areas and dimensions:

3rd floor	– residential	43.2 m²
2nd floor	– residential	37.5 m²
1st floor	– residential	41.1 m²
Ground floor — Sales		39.8 m²
	– Rear Store	10.2 m²
Total		171.8 m²

Overall frontage (including passageway)	4.32	m
Gross frontage	3.98	m
Net frontage	3.2	m
Internal Width (Average)	3.2	m
Shop Depth	22.5	m
Built Depth	31.6	m

> 6.0 Measurements must be strictly in accordance with the RICS Code of Measuring Practice. In the case of let property these areas should be compared to any specified in leases or their schedules/plans, etc. and compared to any areas specified by selling agents. Explanation of retail zones may be necessary.

7.0 Services

We understand that the property is supplied with mains water, electricity and drainage.

> 7.0 Considerably more detail on Services may be needed for other property types where quantity and quality are critical factors.

8.0 Rating

We have made verbal enquiries of XYZ Council and are informed that the property is included within the current Valuation List having the following description and assessments:

Description	Rateable Value (2005)
Shop and Premises	£26,000

> 8.0 This will include details of the current level of Uniform Business Rates and should mention any appeals made or outstanding.

9.0 Town Planning

We have made verbal enquiries of the Local Planning Authority, XYZ Council, and understand that the property is a Grade II Listed Building situated within the City Centre Conservation Area as defined by the City of XYZ Local Plan. The listing is recorded as:

'House now shop. Mid 19th century, circa 1900 shop front grey (yellow) brick; flemish bond, ashlar dressings. Welsh slate roof. Six panel door at right and recessed shop door at left have fan lights with glazing bars. Rounded top light to shop window with similar glazing bars. Shop door has bevelled glass in patterned glazing bars. Tuscan pilasters and bracketed cornice frame shop. Upper floors have sashes with glazing bars. Projecting stone sills cut back underneath at forty five degrees. Chamfered stone quoins support similar cutback in cornice with paved brackets and finishing in pyramidal coped blocks. Low pitched roof with tall banded chimney on left.'

Policy E18 in the Local Plan generally aims to protect and enhance the character, appearance and setting of the City's conservation areas. Policy E19 sets out a number of restrictions and guidelines relating to the City Centre Conservation Area itself, covering the visual historical and architectural importance of the buildings.

Under the section headed 'Land Use Policies' in the Local Plan it is stated that no major development or redevelopment for shopping will be allowed in the City Centre except in the shopping and business centre as allocated in the XYZ Zone (Policy CC1). Policies CC2 and CC3 stipulate that in the Shopping and Business Centre commercial and community purposes will take precedence over other uses except where upper floor space is to be converted to residential use. Furthermore, XXXX is included in the area in which changes of use of ground floors to office use (as covered by Class II of the Town and Country Planning (Use Classes Order) 1972), betting offices and amusement arcades will not be permitted. Elsewhere in the Shopping and Business Centre office users will be given consideration by the Local Planning Authority provided they do not exceed more than 10% of the net frontage length of ground floor properties in retail and commercial uses.

We are informed that there are no plans to review traffic management in the XXXX Place area to restrict service vehicle access during the shopping hours.

We have traced the planning history through the local planners. It has been assumed that the property has planning consent for use as a shop and we are aware of a planning consent dated 11 February 1982 permitting change of use from residential to office of the 1st, 2nd and 3rd floors which expired in 1987. We are informed that this is the only planning consent given since 1974.

A new development is planned for the site on XXXX between Boots and No 10 XXXX which will span the slip road and the through-road off XXXX Bridge. The developers are ABC Ltd and the landowners are the City Council. We anticipate that this development will improve XXXX as a shopping location and increase the pedestrian flow past the subject property.

CDE Ltd and the City Council are planning a joint development of a scheme incorporating a hotel, 25,000 m² of retail, 1,000 space carpark, an ice-rink and a swimming pool, on the site of the existing ice-rink and car park north of A Road by the river. It is anticipated that an application for planning permission will be submitted in 2000 but this proposal is very much in its early stages. However its location should serve to enhance XXXX.

In B Street XXXX Estates are currently refurbishing Nos 68–70 and the ABC Centre on the other side of the river has recently been extended with only one remaining unit unlet. North of the City a mixed development of shops, industrial and residential units is to be constructed near XXXX. An application has been made on a site to the east of the City for an industrial estate incorporating bulky goods retail warehouses.

We are of the opinion that none of the developments or policies described are likely to adversely affect the subject property and we are not aware of any other relevant current policies or proposals.

9.0 This is a critical section and will include details of current policies, existing consents or established uses and will place the property in the context of the area. Details of proposed changes in highways, pedestrianisation schemes and all other relevant issues must be noted, together with the valuer's opinion of potential impact on the property and its future.

10.0 Tenure

The property is freehold with a right-of-way in Fee Simple granted to the occupiers of the adjoining premises over the passageway along the side of the subject property. This gives access to the upper floors of the adjoining building.

10.0 It is increasingly important for valuers to have sight of the title deeds. Where these are not available an appropriate caveat should be included. The danger is that a valuation based upon an assumption of a clean title may subsequently be proved erroneous if the title is in any way restricted.

11.0 Occupation and letting

The entire property was let to XXXX Ltd by virtue of a lease dated 00/00/00 for a term of 25 years from 00/00/00 subject to five-yearly upward only rent reviews at a current rent of £24,000. The lease is drawn on full repairing and insuring terms with the landlord effecting insurance and recovering the premium direct from the tenant.

The lease was recently assigned to . . . . and . . . . The tenant is prohibited from assigning or underletting the whole or part of the demised premises without prior written consent from the landlord which is not to be unreasonably withheld.

The tenant covenants 'not without consent in writing of the lessor first obtained to use the demised premises or any part thereof or suffer the same to be used otherwise than as a high-class retail shop only'. Although the upper floors are fitted out for residential use and such a use would be acceptable in planning terms, we have valued the upper parts as storage ancillary to the retail in accordance with the terms of the lease.

11.0 This section should specify whether leases or underleases have been inspected by the valuer or whether reliance is being placed on information provided by solicitors. Normally the purchaser's attention is drawn here to key covenants on repairs, insurance, alterations, alienation, user, rent reviews and service charges.

12.0 Covenant

We have no information on the covenant of the tenant and have assumed that you are satisfied with the covenant status.

We understand from your solicitors that the assignment does not appear to be documented and they should satisfy themselves as to the legal position in this regard, particularly as we understand that a schedule of repairs dated 00/00/00 is still outstanding and should have been completed within six months of the date of the licence to assign.

> 12.0 Valuers may be asked to undertake company searches or credit rating enquiries. Tenant covenant is a key factor in today's market and both valuer and purchaser need to know the quality of the covenant.

13.0 Rental value

XXXX is a Cathedral City with a substantial catchment area and is much sought after by national multiple retailers. The main shopping area in the City Centre is centred around A Street and XXXX with B Street being relatively secondary in terms of rental value and location.

Details of all current lettings known to us are set out in Appendix 5 together with our analysis and conclusions.

The current rent reserved under the lease is £24,000 a year.

We are of the opinion that the current market rental value of the property is approximately £30,000 a year, equating to £400m^2 Zone A.

> 13.0 In an investment valuation report the purchaser should be drawn to the same conclusion or opinion as the valuer. Here the valuer's opinion of current market rental value and of market capitalisation rate or yields should be made, supported by market data. Where market data is thin this must, as here in 15.0, be drawn to the purchaser's attention.
>
> Purchasers also expect some guidance on anticipated market movements.
>
> Are rents rising? If so, is this supported by evidence of recent lettings.

14.0 Valuation

Terms have been agreed to purchase this freehold shop investment in the sum of £525,000 (five hundred and twenty five thousand pounds) subject to contract.

We therefore estimate that the agreed acquisition price reflects a net initial yield of 4.32% rising to 5.40% on reversion in 00/00 to produce a net equivalent yield of 5.27%. These yields are after allowance for acquisition costs of 5.75% to cover fees and Stamp Duty.

> 14.0 The value must be expressed in words. Where evidence is available the valuer should indicate the basis for the capitalisation rates used.

15.0 Recommendation

We draw attention to the fact that at the time of valuation there was a limited amount of directly comparable rental evidence available on which to base our valuation, particularly taking account of the limited shop width. It is relevant that we were aware of other parties bidding for the investment and indeed, having submitted an offer for this property we were informed that two other offers had been received by the vendor at the same level and are subsequently informed of a further offer in excess of the agreed contract price.

We are of the opinion that the agreed purchase price of £525,000 (five hundred and twenty five thousand pounds) for this freehold shop investment is appropriate in the current market and recommend the purchase to . . . . Ltd.

> 15.0 This section may be split between the recommendation — to buy at — and the investment considerations. The latter would develop the valuer's opinion of the state of the market supported by direct evidence, research findings and observations based on, say, the IPD forecasts and an indication of possible growth expectations and yields given that growth.

16.0 Limitations

Any limitations on the valuation must form part of the terms of engagement.

The minimum terms of engagement are set out in the RICS Appraisal and Valuation Standards under PS21 with a more detailed commentary under Appendix 2.1,

The minimum terms must cover:

(a)   identification of the client
(b)   the purpose of the valuation
(c)   the subject of the valuation
(d)   the interest to be valued
(e)   the type of property and how it is used, or classified, by the client
(f)   the basis, or bases, of valuation
(g)   the date of valuation
(h)   the status of the member and disclosure of any previous engagement
(i)   where appropriate the currency to be adopted
(j)   any assumptions, special assumptions, reservations, any special instruction on departure
(k)   the extent of the members' investigations

(l)   the nature and source of information to be relied on by the member
(m) the requirement of consent to publications
(n)   any limits or exclusions of liability to parties other than the client
(o)   confirmation that the valuation will be undertaken in accordance with these (Red Book) standards
(p)   the basis on which the fee will be calculated
(q)   the members', or organisation's complaints handling procedure, with a copy available on request.

Specific limitation under j, m, n will usually be repeated in the report and would typically cover: (under Assumptions Red Book Appendix 2.2):

- title
- condition of building — a valuation is not a building survey
- services — services are not normally tested so the assumption is that the services, and any associated controls or software, are in working order or free from defect
- planning
- contamination and hazardous materials
  - the valuer must state the limits on the investigations in these areas
- environmental
  - the valuer should state the limits that will apply to the extent of the investigations.

In addition some Professional Indemnity Policies may require specific limitation statements to be included in all valuation reports.

Appendices

- Terms of engagement
- Definition of Market Value
- Plans
- Photographs
- Details of Market Analysis

17.0 Signature

Additional information which may be requested by some investor clients

When advising investors on specific property investment opportunities it is normal practice, in agreement with the client, to provide analysis which goes beyond the basic information on initial, reversion and equivalent yields. The following section describes the nature of the analysis that might be requested and outlines/examples of the related calculations. The following case is based on a 2005 transaction, but details have been changed.

Background

An institutional client holding a large portfolio is seeking to achieve growth in its revenues, which are almost wholly in the form of rents, in excess of inflation. It must certainly achieve return in excess of the redemption yield on government bonds, which are risk free and liquid.

It has significant cash to invest, is seeking to invest in non-traditional (offices and retail) sectors and the market is very competitive.

The investment

It has been offered an investment for an asking price of £29m. The subject property is a 41-acre specialist storage site which a subsidiary of a well known global brand has leased for a remaining 15 years with rents rising at a fixed 3% pa. Last year's rent was £1.85m.

Due to its location the site is particularly suitable for the existing tenant, and if the existing tenant were not in place then it would be difficult to imagine re-letting the land for the same purpose. The alternative use would be as general storage, for which it would be worth around £25,000 per acre or £1.025m pa, or as building land for industrial units, for which it could be sold for around £230,000 per acre or £9.43m.

The total outlay will be £29m plus stamp duty at 4%, agents' fees at 0.35%, legal fees at 0.25% and VAT on fees at 17.5%. This sums to (a rounded) £32.2m.

The advice

Consultant surveyors were employed to advise on the purchase. A summary of their recommendations is as follows.

First, the consultants outline the 'running yield profile' of the asset by showing the relationship between the rent received in any year and the purchase price. This is as follows

Year	Rent	Yield
1	£1.9627	6.10%
2	£2.0215	6.28%
3	£2.0822	6.47%
4	£2.1447	6.66%
5	£2.2090	6.86%
6	£2.2753	7.07%
7	£2.3435	7.28%
8	£2.4138	7.50%
9	£2.4862	7.72%
10	£2.5608	7.95%
11	£2.6377	8.19%
12	£2.7168	8.44%
13	£2.7983	8.69%
14	£2.8822	8.95%
15	£2.9687	9.22%

The yield on reversion was expressed as 9.22%, with the reversionary yield beyond the 15 year point in 2020 dependent on whether the existing lessee renews the lease or the value reverts to industrial values.

If the tenant does not renew the lease, the yield on reversion yield is £1.025m/£32.2m = 3.2%. Note that this measure uses the current rental value, and not a projection of the rental value of the land in year 15. This fixation on current values gives the property market participant an initial feel for the nature of the investment: 6.1% rising to 9.2% over 15 years falling back to 3.2% thereafter describes a particular pattern of risk and return which may be compared with a more straightforward fully leased investment which yields 6% flat. The yield pattern described in the table above is misleading, as it ignored the more likely yield on reversion.

The consultants then analyse the investment in terms of IRR. A whole series of questions then arise.

- What holding period should be adopted?
- What assumptions should be used?

15 years lends itself readily as the appropriate period, but five and ten will also be of interest. Then an assumption has to be made regarding the reversion. Will the tenant renew his lease? At what rent and under what terms will he renew?

Will the tenant vacate? If he vacates, the land may be sold as building land for industrial units, for which it could be sold currently for around £230,000 per acre or £9.43m. What will this value be in 15 years?

If the tenant renews, for what time period will we now project the cash flow? For simplicity, and in accordance with market practice, it may be best to assume a sale of the investment at year 15 following the re-letting. This comes back to comparable evidence. What initial or all risks yield would a purchaser be likely to accept for the property if it has been let for an undefined period at a rent rising at 3% pa?

Assuming a re-letting at the year 15 rent with continued 3% uplifts, the reversion is valued at an 'exit yield' of 6.75% (there is limited evidence for this). The reversion can then be sold for £2.9687m capitalised at 6.75%.

£2.9687/.0675 = £43.98m.

Assuming annual in arrears cash flows for simplicity, the cash flow then looks like this:

Year	Capital	Rent	Total
0	–£32.20		–£32.20
1		£1.96	£1.96
2		£2.02	£2.02
3		£2.08	£2.08
4		£2.14	£2.14
5		£2.21	£2.21
6		£2.28	£2.28
7		£2.34	£2.34
8		£2.41	£2.41
9		£2.49	£2.49
10		£2.56	£2.56
11		£2.64	£2.64
12		£2.72	£2.72
13		£2.80	£2.80
14		£2.88	£2.88
15	£43.98	£2.97	£46.95

Using the Excel IRR function, this is an IRR of 8.52%. Assuming no-re-letting and a sale of the land as building land, we need to make an assumption about the change in land values over this period. Land values have increased at something close to the rate of inflation in the long run, but in this case the consultant has assumed that land has increased at rates varying from 2% to 6%. At 2%, we get:

Year	Capital	Rent	Total
0	−£32.20		−£32.20
1		£1.96	£1.96
2		£2.02	£2.02
3		£2.08	£2.08
4		£2.14	£2.14
5		£2.21	£2.21
6		£2.28	£2.28
7		£2.34	£2.34
8		£2.41	£2.41
9		£2.49	£2.49
10		£2.56	£2.56
11		£2.64	£2.64
12		£2.72	£2.72
13		£2.80	£2.80
14		£2.88	£2.88
15	£12.69	£2.97	£15.66
		IRR	4.45%

The IRR rises to 5.19% assuming 4% growth in land values and 6.04% assuming 6% growth.

Looking at the investment over five and ten year holding periods, the difficulty which arises is the following question: at what price could the investment be sold at the end of the holding period?

In answering this question, the consultants split the cash flow at reversion into two parts. At year 5, the following year's cash flow of £2.28m was split into the underlying rental value of £25,000 per acre, with rental growth of 4% applied (see above), and a remaining 'froth' or high risk element dependent upon continued occupation by the existing tenant.

The rent components were as follows:

£25,000 * 41 = £1.025m total rental value.

£1.025m * (1.04)^5 = £1.247m. This is the ERV component of rent.

The 'froth' component of rent is then £2.21m − £1.247m = £0.963m.

Different exit yields were then applied to each component. 6.69% was applied to the ERV and 8% to the froth.

The exit value is then given by the following:

£1.247m/.0669	=	£18.64m
£0.963m/.08	=	£12.04m
Total: £18.64 + £12.91m	=	£30.68m

The five year cash flow then looks like this:

Year	Capital	Rent	Total
0	−£32.20		−£32.20
1		£1.96	£1.96
2		£2.02	£2.02
3		£2.08	£2.08
4		£2.14	£2.14
5	£30.68	£2.21	£33.76

The IRR is 5.6%.

A similar analysis for a 10 year holding period uses slightly higher exit yields as there is increasing risk as the lease end approaches. The IRR which results is around 6.75%.

Modelling risk and return

The 5, 10 and 15 year gilt redemption yields are all around 4.5%. Does a range of potential returns of (roughly) 4.5% to 8.5% provide adequate compensation for risk?

- Have all permutations been considered? It would be helpful to see an analysis of the likely return if the lease is renegotiated on reasonable terms (what are they likely to be?) and more variations regarding the planning position and alternative use values.
- What are the relative probabilities of the higher and lower return outcomes? The permutations are many, and a simulation may well be the best approach to this problem.

A further set of issues arises.

- Does the conventional valuation format using running yields add any value?
- What evidence does the valuer/consultant have for his choice of yields?

Explicit cash flow modelling is superior, and the inadequacy of the traditional valuation approach is illustrated very well in this case. However, the need to establish an exit yield remains.

Appendix C

Illustrative Development Site Appraisal Report

The following report is based on the standard report produced by Circle Systems Development Appraisal System using the data from Example 11.1 on p215.

The report is annotated with comments to explain the functionality of the system and the additional variables not analysed in the simple illustrative example in Chapter 11.

Contents page
Circle Version 2.00

Section 1

TIMESCALE & SUMMARY
PROJECT SUMMARY

CONTENT
Part 1

	m²	
	Gross	Net
Offices	7,000	6,000
Total	7,000	6,000

Net: Gross Ration = 85.71

TIMESCALE
Part 1

Commencement	Oct 1995
Pre-Construction Void	0 months
Construction Start	Oct 1995
Construction Duration	24 months
Void begins	Oct 1997
Void duration	0 months
Terminating end of	Sep 1997
Part timescale	24 months

The first part of the timescale and summary section shows the project summary.

The development may be split into a number of phases or 'parts' each of which would be described in this section.

The project summary shows the gross and net or lettable areas of each area of development and indicates the net to gross efficiency ratio.

The page then shows all of the various time periods that will be used in the calculation.

In our simple calculation there is no void period; this however is unrealistic unless the development is pre-let.

In most cases a void period, ie the period between completion of construction and letting, will be assumed. This period will extend the pay-back period of the funding required for development. It is one of the variables carefully considered when development scenarios are being modelled using this type of software.

Because the void extends the funding of the whole outstanding debt an increase in this period can have a significant impact upon the profitability of a scheme.

FINANCIAL SUMMARY

Project Timescale	24 months	
INCOME		
Sales Income		0
Annual Rental Income	720,000 pa	
Net Cap Value		10,285,714
Other Income		0
Completed Value		10,285,714
EXPENDITURE		
Site Purchase Cost	3,516,795	
Site Purchase Fees	105,504	
Total Purchase Cost	3,622,299	
Construction Costs	2,800,000	
Construction Fees	350,000	
Total Construction	3,150,000	
Sale Costs/Fees	102,857	
Purchaser's Costs	0	
Other Costs	0	
Marketing/Letting	194,000	
Site Finance	1,168,191	
Constr. Finance	505,510	
Void Finance	0	
Offset Interest	0	
Total Finance	1,673,701	
Net VAT	0	
Total Expenditure		8,742,857
Balance		1,542,857
Profit on GDV	15.0%	
Void Interest Cover	1.16 yrs	
Void Rent Cover	2.14 yrs	
Dev Yield on Cost	8.24%	

ASSUMPTIONS

- Construction Interest Compound: Weighted 50%
- Payments annually in advance
- Profit measured against gross development value
- Purchaser's costs based on gross capitalisation
- Fees on Sales based on Sales plus NDV
- Professional Fees based on construction excluding demolition and site works

This page gives a financial summary of the value of the development and lists the development expenditure.

The system can be used for capital sales (eg residential development) and capitalised rental income or a mixture of the two. The expenditure summary shows additional items not considered in the simple manual calculation on p000. For example VAT and offset interest.

This page also shows some useful statistics for the development scheme.

The Profit on Gross Development Value (or costs) as selected by the user.

In this case the profit has been stipulated by the user at 15% in order to calculate site value. Where the system is used for development appraisal and the land value is known: it is this figure which will be calculated as the residual sum.

The void interest cover shows the period of time over which the profit would be completely eroded by continuing interest payments where the development remains un-let.

The void rent cover is calculated by dividing the residual profit by the rental value. If the developer is guaranteeing rent as part of a funding arrangement a profit will still be made providing the development is let within the period of years shown, in this case 2.14 years.

The Development Yield on cost is calculated by dividing the income by the total costs and expresses income as a percentage of total costs incurred in creating the scheme. The assumptions of the calculation including the weighting of the interest on construction costs (in this case 50%) are clearly shown on this page of the report.

Circle Version 2.00   **Section 2**
APPRAISAL

	%		
REALISATION	%		
SALES	N/A		
GROSS RENT 6000 m² @ 120.0 m²	720,000 pa		
Less Ground Rent	0 pa		
Net Rental Income	720,000 pa		
CAPITALISATION @ Yield 7.00% × 14.29 YP		10,285,714	
(NDV = 10,285,714)		GDV	10,285,714
Other Revenue		0	
NET REALISATION			10,285,714
OUTLAY			
SITE PURCHASE PRICE			3,516,795
Stamp Duty 1.000%	35,168		
Land Acq Agent 1.000%	35,168		
Land Acq Legal 1.000%	35,168		
Grd Rent Agt Fee	0		
Grd Rent Lgl Fee	0		
Town Plan/Survey	0		
Arrangement Fee	0	3,622,299	
CONSTR. 7000 m² @ 400.00 m²	2,800,000		
Contingency	0		
Demolition	0		
Site Works/Roads	0		
FEFS			
Architect 9.000%	252,000		
Quant Surveyor 3.5000%	98,000		
Struct Engineer	0		
Mech/Elec Eng	0		
Misc Fees	0		
Project Manager	0		
Statutory costs	0		
Misc Costs	0		
MARKETING	50,000		
LETTING Agt/Legal 20,000%	144,000	3,344,000	
Purchasers Costs	0		
Sale Agent Fees 1.000%	102,857		
Sale Legal Fees	0	102,857	
EXTRA COSTS	0		
Value Added Tax	0	0	
INTEREST (site 15.00% building 15.00%)			
Site (excl void) 24 mth	1,168,191		
Building (Comp 50%) 24 mth	505,510		
Void 0 mth	0		
Offset Interest	0		1,673,701
COSTS			8,742,857
PROFIT 15.00%			**1,542,857**

This one page appraisal shows clearly the key components of the appraisal. Where a development is phased the results of each phase are aggregated into the single appraisal.

The layout of the appraisal is the same whether you are using it for calculating the residual land value or residual profit.

In this case the land value is calculated and inserted as the sale purchase price in the outlay section. The appraisal is calculated at the pre-determined target profit rate of 15%.

The contingency is a figure, usually a percentage of construction costs, to reflect the difficulties of precisely estimating building costs and to cover extras in the contract that may occur for unforeseen circumstances.

The system can cope with a wide range of professional fees, VAT and extra income and expenditure which can be inserted using a number of expandable screens from within the main appraisal screens.

The difference in figures between the Circle analysis and the manual calculation on page 122 is due to the more realistic treatment of marketing costs based on a percentage of the first year's rent and a more accurate calculation of the interest charges.

Circle Version 2.00   **Date 07/11/1995**

SENSITIVITY ANALYSIS

Residual Land Value Table

	Bldg Rate				
	380.00	**390.00**	**400.00**	**410.00**	**420.00**
Rent Rate					
105.00	2,871,328	2,804,247	2,737,160	2,670,070	2,602,984
112.50	3,261,153	3,194,069	3,126,980	3,059,890	2,992,801
120.00	3,650,976	3,583,884	3,516,795	3,449,708	3,382,616
127.50	4,040,793	3,973,699	3,906,614	3,839,521	3,772,432
135.00	4,420,608	4,363,519	4,296,432	4,229,340	4,162,246

The system can be used to generate a sensitivity analysis matrix.

In this case the matrix shows the effect on the residual land value of changes in two variables : the rent and the building cost rates.

The 'worst' case scenario is therefore in the top right-hand corner of the matrix and the 'best' case scenario in the bottom left-hand corner.

When using the system for appraisals the matrix would show the amount of profit in £ (or other currency) and as a percentage of either GDV, NDV or construction costs as pre selected by the user.

The sensitivity analysis can also produce a three-variable matrix by using five layers or windows which are printed separately.

Appendix D

Solutions to Questions Set in the Text

Solutions to Part I questions

1   £975.33 (A£1)
2   £663.429 (PV£1)
3   £3,553.20 (Annuity × or ÷ PV £1 pa)
4   £21,321.44 (A£1 pa)
5   £97.38 (ASF)
6   (a)   Lump sum or annual sinking fund
      (b)   £8,049.25 or £1,424.60 pa
      (c)   £25,000 × A £1 at 12% = £25,000 × 3.1058 = £77,645 × PV £1 at 12% = £25,000
             Therefore £25,000 or £77,645 × ASF 0.05698 = £4,424.21
7   (a)   £2,763 (A £1 pa)
      (b)   £5,764.75
      (c)   £8,085.36
8   £465,292 (£10,000 × PV £1 pa at 8% +£ 50,000 × PV £1 pa in perp at 8% × PV £1 in 10 years at 8%)
9   13%

Solutions to Part II questions

Chapter 5

Q   Yields

$$\frac{75,000}{1,250,000} \times 100 = 6\% \text{ (All Risks Yield/ARY or capitalisation (cap) rate}$$

Note: $\dfrac{1,250,000}{75,000} = 16.667$ Years' Purchase $= \dfrac{100}{16.667} = 6\%$

Q   Equivalent Yield Analysis

Term Income	£30,000		
YP 2 years at $x\%$	(1.8080)	(£54,240)	
Reversion Income	£40,000		
YP per at $x\%$			
PV £1 in 2 years at $x\%$	(12.477)	(£499,080)	(£553,320)
			£553,000

By trial and error the rate percent that discounts the income to a present value of £553,000 will be found to be 7% to the nearest per cent. This is the equivalent yield, which in turn is the internal rate of return.

In both questions valuers might deduct management fees from the net re before seeking to calculate the capitalisation rate or ARY.

Chapter 6

Growth Rate

ARY(k)	=	6%
Gilts	=	8%
Target rate (e)	=	8% + 1% = 9%
Rent review (t)	=	5 years
Substituting in		

$$k = e - (\text{ASF} \times \text{P})$$

0.16709P	=	0.09 − 0.06
0.16709P	=	0.03
P	=	0.03/0.16709
P	=	0.18 (18% over 5 years)

Therefore annual growth is:

$(1 + g)^5$	=	1.18
$(1 + g)$	=	$\sqrt[5]{1.18}$
or $(1 + g)$	=	$1.18^{0.2}$
$(1 + g)$	=	1.03365
g	=	1.03365 − 1
g	=	0.03365 × 100
	=	3.365% per year

Q   Freehold Warehouse

Step 1: Find the ARY from comparable

$$\frac{£\,42,500}{£472,000} \times 100 \quad = \quad 9.0\%$$

Step 2: Find rental value from comparable
Given as £42,500

Step 3: Equivalent Yield Valuation
Assume no market variation from ARY

Term Income	£30,000		
YP 2 years at 9%	1.7591		£52,773
Reversion Income	£42,500		
YP perp def'd 2 yrs at 9%	9,3520		£397,460
			£450,233

Step 4: Modified DCF — Rational Method

K	=	0.09
e	=	0.12
t	=	5-year standard review pattern
K	=	$e - (ASF \times P)$
0.09	=	$0.12 - (0.15741P)$
0.15841P	=	$0.12 - 0.09$
0.15741P	=	0.03
P	=	0.03/0.15741
P	=	0.19058 (say 0.19)

Therefore annual growth

$(1 + g)^5$	=	0.19 (Note: A £1 formula is $(1 + i)^n$ is $1 +$ interest to produce £1.19 in 5 years)
$(1 + g)$	=	$\sqrt[5]{1.19}$
or $(1 + g)$	=	$1.19^{0.2}$
$(1 + g)$	=	1.0354
g	=	0.0354×100
	=	3.54%

Term Income	£30,000		
YP 2 years at 12%	1.6901		£50,703
Reversion Income	£42,500		
A £1 in 2 yrs at 3.54%	1.072		
	£45,562		
YP perp at 9% 11.111			
PV £1 in 2yrs at 12% 0.79719	8.8575		£403,571
			£454,274

Step 5: Real Value Approach
Assess IRFY

$$\frac{e - g}{1 + g}$$

$$\frac{0.12 - 0.0354}{1.0354}$$

$$\frac{0.0846}{1.0354}$$

$0.0817 = 8.17\%$

Term Income	£30,000		
YP 2 yrs at 12%	1.6901	£50,703	
Reversion Income	£42,500		
YP perp at 9% 11.111			
PV £1 in 2yrs at 8.17% 0.8546	9.4959	£403,578	
		£454,281	

(Note: a number of calculations have been rounded giving rise to a marginal difference in totals between Rational and Real Value figures.)

   In this example the variation in opinion of value is small and in valuation terms the opinion might be rounded in all cases to £450,000. The variation is small due to a number of factors; namely, the short period to the reversion, the relatively low implied growth rate and the fact that the term income is some 70% of the market rental value.

Chapter 7

Q   Leasehold Valuation

Profit rent	£20,000	
YP 4 years at 8% and 3% Tax 40%	2.0904	
	£41,808	
Capital Value	£41,808	
Net Income	£12,000	

(£20,000 less tax at 40%)

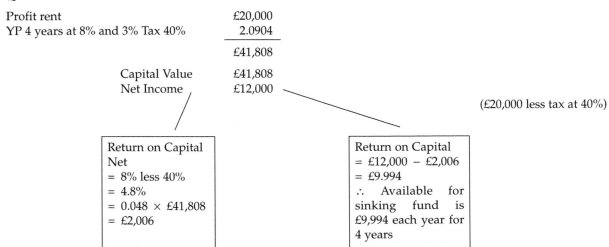

Return on Capital
Net
= 8% less 40%
= 4.8%
= 0.048 × £41,808
= £2,006

Return on Capital
= £12,000 − £2,006
= £9.994
∴   Available   for
sinking   fund   is
£9,994 each year for
4 years

Sinking fund	=	£9,994
A £1 pa for 4 years at 3%	=	4,1836
		£41,810

Note small cumulative error flowing from YP of 2.0904 instead of 2.0903738

Q   Shopping Centre — geared leasehold

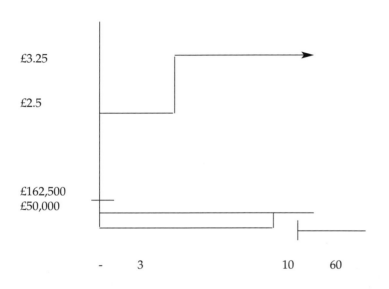

Conventional
Profit Rent Structure

Conventional Valuation of 60-year lease

Current Rent Receivable		£2,500,000	
Head Rent		50,000	
Profit Rent		£2,450,000	
YP 3 years at 8% and 3% tax 40p		1.6149	£3,956,505
Reversion to		£3,250,000	
Head Rent		50,000	
Profit Rent		£3,200,000	
YP 7 years at 8% and 3% tax 40p	3.3612		
PV £1 in 3 years at 8%	0.79383	2.6682	£8,538,308
Reversion to		£3,250,000	
Head Rent (5%)		162,500	
Profit Rent		£3,087,500	
YP 50 years at 8% and 3% tax 40p	10.5512		
PV £1 in 10 years at 8%	0.46319	4.8872	£15,089.261
			£27,584.074

Notes
- Variable profit produces 3 separate sinking funds, one for each tranche of income when only one is needed to recover capital over 60 years which is, in valuation terms, approaching perpetuity.
- Valuation should be reworked using Sinking Fund method to correct arithmetic error.
- Conventional approach fails to reflect the difference between fixed profit rents and growth profit rents.
- Conventional approach cannot adequately deal with the profit rent growth from year 10 to year 60.
- Head rent review in year 10 is taken to be 5% of todays estimate of open market rental value.

All these issues have probably resulted in an undervaluation of the leasehold interest.

DCF Valuation

Step 1: Calculate the implied growth rate. Using market yields requires the valuer to adopt freehold yields.

Given

K	=	0.07 (conventionally freehold is 1–2% below leasehold)
e	=	0.10
P	=	unknown
t	=	5 years

Then

K	=	$e - (\text{ASF} \times P)$
0.07	=	$0.10 - (0.1638\text{OP})$
0.1638OP	=	$0.10 - 0.07$
0.1638OP	=	0.03
P	=	$0.03/0.16380$
P	=	0.18315
P	=	18.315% (over 5 years)
$\therefore$ P	=	$(1 + g)^t - 1$
1.18315	=	$(1 + g)^5$
$\sqrt[5]{1.18315}$	=	$(1 + g)$ or $1.18315^{0.2}$
1.0342-1	=	g
0.0342	=	g
3.42%	=	g

Assuming income is all annual in arrears, and allowing for implied rental growth in the cash flow at 3.42% on the OMRV of £3.25m, produces the following cash flow which is discounted at a net of tax rate on next of tax income.

Period (end of year)	Shop Rents	Head Rent	Profit Rent at 40p	Net of Tax at 7.2%[1]	PV £1 pa at 7.2%	PV £1	PV
1	2,500,000	50,000	2,450,000	1,470,000	–	0.9328	1,371,216
2	2,500,000	50,000	2,450,000	1,470,000	–	0.8702	1,279,194
3	2,500,000	50,000	2,450,000	1,470,000	–	0.8117	1,193,199
4[2]	3,594,983	50,000	3,544,983	2,126,989	–	0.7572	1,610,556
5	3,594,983	50,000	3,544,983	2,126,989	–	0.7063	1,502,292
6	3,594,983	50,000	3,544,983	2,126,989	–	0.6589	1,401,473
7	3,594,983	50,000	3,544,983	2,126,989	–	0.6146	1,307,247
8	3,594,983	50,000	3,544,983	2,126,989	–	0.5734	1,219,615
9[3]	4,253,225	50,000	4,203,225	2,521,935	–	0.5348	1,348,731
10	4,253,225	50,000	4,203,225	2,521,935	—	0.4989	1,258,193
11[4]	4,253,225	212,661	4,040,564	2,424,338	–	0.4654	1,128,287
12	4,253,225	212,661	4,040,564	2,424,338	–	0.4342	1,052,647
13	4,253,225	212,661	4,040,564	2,424,338	–	0.4050	981,857
14–18[5]	5,031,991	212,661	4,819,330	2,891,598	4.078	0.4050	4,775,734
19–23	5,953,348	212,661	5,740,687	3,444,412	4.078	0.2860	4,017,245
24–28	7,043,406	212,661	6,830,745	4,098,447	4.078	0.2021	3,377,791
29–33	8,333,054	212,661	8,120,393	4,872,235	4.078	0.1427	2,835,302
34–38	9,858,837	212,661	9,646,176	5,787,705	4.078	0.1008	2,379,108
39–43	11,663,990	212,661	11,451,329	6,870,797	4.078	0.0712	1,994,961
44–48	13,799,966	212,661	13,587,305	8,152,383	4.078	0.0503	1672244
49–53	16,326,739	212,661	16,114,078	9,668,446	4.078	0.0355	1,399,695
54–58	19,316,165	212,661	19,103,504	11,462,102	4.078	0.0251	1,173,235
59–60	22,852,955	212,661	22,640,294	13,584,176	1.803	0.0177	433,513

£40,713,331

1. 12% gross less tax at 40p = 7.2% net.
2. £3,250,000 (ERV) × A £1 in 3 years at 3.42% = £3,594,983.
3. Shop rents reviewed every 5 years with implied growth 3.42%.
4. Head rent reviewed at 5% of shop rents, taken here as 5% of rents collected but could in some leases be 5% of OMRV. Head rent remains fixed for 50 years.
5. Each 5-year cash flow is discounted for five years using the PV £1 pa factor. The resultant sum is the present worth at the beginning of the 5 years and which must then be discounted to point zero using the PV £1. In the period 14–18 the PV £1 pa has brought the cash flow back to the end of year 13 (see formula for PV £1 pa) and hence the use of the PV for 13 years not 14 years.
6. This opinion of value is £13m more than that arising from a conventional dual rate approach. The errors in the dual rate, if corrected would reduce this difference but the critical factor is that the conventional approach only partially allows for the profit rent growth from year 10 by using a remunerative rate of 8%. The DCF using an implied growth rate identifies this investment opportunity. The issue for the valuer is one of determining (open) market value and unless the DCF approach can be shown to be the market method then the conventional method may represent the best price one can expect in the market place.

Chapter 9

Q Premium Calculation

Freehold All Risks Yield	=	5%
Leasehold	=	6%/3% tax 40p

Then:

Proposed Profit rent or rental loss is:		£75,000
less		£50,000
		£25,000

Average of YP 5 years at 5%	4.3295	
and YP 5 years at 6%/3% tax 40p	2.6743	3.502
Premium		£87,547

1. In theory there could be no agreement using this conventional approach. Comparing the Freehold YP of 4.3295 and the Leasehold YP of 2.6743 indicates that the freeholder needs a larger premium than the tenant can afford.
2. In practice the freeholder is likely, under normal market conditions, to insist on a sum close to £25,000 × 4.3295; that is, £108,237.

Q Surrender and Renewal

(a)

A Freeholder's Present Interest

Current Rent	£10,000	
YP 10 years at 9%	6.4177	64,177
Reversion to MRV	£100,000	
YP perp defíd 10 years at 9%	4,6935	469,350
		£533,527

Equivalent yield approach. The ARY of 8% has been revised by 1% to reflect the fixed nature of the rent for the next 10 years. A rational or real value approach should be used as a check.

B Freeholder's Proposed Interest

	£x	
Rent to be reserved		
YP 5 years at 8%	3.9927	3.9927x
Reversion to	£100,000	
YP perp defíd 5 years at 8%	8.5073	850,730
		£850,730 + 3.9927x

Let:

Value of Present Interest	=	Value of Proposed Interest
£533,527	=	£850,730 + 3.9927x
£533,527 − 850,730	=	3.9927x
−£317,203	=	3.9927x
−317,203/3.9927	=	x
−£79,445	=	x

This transaction is sufficiently beneficial for the freeholder to be able to pay £79,445 a year to the tenant for the next five years.

C   Tenant's Present Interest

Market Rental Value	£100,000
Rent Payable	10,000
Profit rent	£90,000
YP 10 years at 10%/3% tax 40p	4.0752
	£366,768

D   Tenant's Proposed Interest

Market Rental Value	£100,000
Rent to be reserved	x
Profit rent	£100,000 − x
YP 5 years at 10%/3% tax 40p	2.4159
	£241,590 − 2.4159x

Let:

Value of Present Interest	=	Value of Proposed Interest
£366,768	=	£241,590 − 2.4159x
£366,768 − £241,590	=	−2.4159x
£125,178/2.4159	=	−x
£51,814	=	−x

This suggests, because of the dual rate approach, that the tenant would be willing to accept £51,814 (say £52,000). On these terms a deal would be negotiated and both parties would be financially better off.

(b)

A   Freeholder's Present Interest as in (a) = £533,527

B   Freeholder's Proposed Interest

Open Market Rental Value	£100,000	
YP perp at 8%	12.5	£1.250,000
Less reverse premium to tenant		x
		£1,250,000 − x

Present Interest	=	Proposed Interest
£533,527	=	£1,250,000 − x
x	=	£1,250,000 − £533,527
x	=	£716,473

C   Tenant's Present Interest as in (a) = £366,768

D   Tenant's Proposed Interest

Market Rental Value	£100,000
Rent to be paid	£100,000
Profit rent	£0
Tenant's Proposed Interest =	£0

Here the tenant needs £366,768 and the freeholder will pay up to £716,473 to secure the surrender; hence a negotiated settlement will be achieved.

These calculations illustrate how important it is to assess the strengths and weaknesses of both parties before commencing negotiations. It also illustrates how the use of a dual rate year's purchase places a different 'value' on what is essentially the same monetary gain or loss.

This problem could be reset as a marriage value exercise. The freehold value in possession is £1,250,000, while the freehold subject to the lease is £533,527 and the leasehold is £366,768. The combined value is £900,295 and hence a marriage or merger value of £349,705 exists (£1,250,000 − £900,295).

Q   Non Standard Rent Review

The formula approach as set out on p170 produces a multiplier k to correct the known rent to the unknown rent thus:

$$k = \frac{(1 + r)^n - (1 + g)^n}{(1 + r)^n - 1} \times \frac{(1 + r)^t - 1}{(1 + r)^t - (1 + g)^t}$$

and

k	=	multiplier
r	=	equated yield of 10% = 0.10
n	=	number of years to review in subject lease = 3
g	=	annual rental growth = 0.05 (this might have to be calculated using the implied growth rate formula)
t	=	number of years to review normally agreed = 5

hence substituting:

$$\frac{(1 + 0.10)^3 - (1 + 0.05)^3}{(1 + 0.10)^3 - 1} \times \frac{(1 + 0.10)^5 - 1}{(1 + 0.10)^5 - (1 + 0.05)^5}$$

$$\frac{1.331 - 1.1576}{1.331 - 1} \times \frac{1.6105 - 1}{1.6105 - 1.2763}$$

$$\frac{0.1734}{0.331} \times \frac{0.6105}{0.3342}$$

$$0.5238 \times 1.8267$$

$$0.9568$$

Therefore:

£25,000 × 0.9568 = £23,920

A landlord in a 5% a year rising market will be willing to accept either £23,920, with rent reviews every three years, or £25,000 when rent reviews are less frequent at 5-year intervals. In theory the position is reversed in a falling market with landlords willing to accept less rent for the security of a longr term, but theory here is distorted by upward-only rent review clauses.

Chapter 10

(a) Full rental value of the building as improved

4th floor	250m² × 200	=	£50,000
Ground, 1st, 2nd, 3rd floors	4 × 400m² × 200	=	£320,000
			£370,000

Section 34 market rent disregarding improvements:

1st, 2nd, 3rd floors	3 × 400m² × 150	=	£180,000
	400m² × 200	=	£80,000
			£260,000

Probable section 34 rent in 3 years on current values: £260,000

(b) The freehold interest

Law of Property Act 1969 — 21-year rule will apply:

Landlord's present interest:			
Current net income		£150,000	
YP 3 years at 7%		2.6243	£393,645
Reversion to S.34 rent		£260,000	
YP 15yrs at 7%	9.1079		
PV £1 in 3yrs at7%	0.81630	7.4347	£1,933,042
Reversion to MRV		£370,000	
YP perp defíd 18yrs at 7%		4.2266	£1,563,842
			£3,890,529

(The courts might allow a review after 11 years to MRV but it has been taken here for 15 years with the assumption of reviews after 5 and 10 years only. ARY reflects market growth expectations in rents and the probability that rent review convenants will protect the tenant with section 34 and LPA 21-year rule disregards.)

Tenant's present interest:		
ERV	£370,000	
Rent reserved	£150,000	
Profit rent	£220,000	
YP 3 years at 8%/3% 40p	1.6149	£355,278

Reversion		£370,000	
Rent reserved (section 34)		£260,000	
Profit rent		£110,000	
YP 15yrs at 8%/3% 40p	5.8458		
PV £1 in 3yrs at 8%	0.79383	4.6803	£514,829
			£870,107

(Variable profit rent needs to be checked with more accurate method.)

If the tenant acquires the freehold, the interests would be merged and would be worth:

	£370,000
YP perp @ 7%	14.2857
	£5,285,714

A freeholder recognising the marriage value will ask for a sum in excess of the open market value of £3,890,529 and would generally require a 50% share of the marriage value of £525,078 (£5,285,714 less the freehold and leasehold values of £3,890,529 + £870,107). Valuing the leasehold dual rate without a tax adjustment increases the value of the leasehold to £1,197,970 and reduces the marriage value to £197,215.

(c) Surrender for new 20-year lase with rent review after 5 years to MRV
Note: Surrender implies the releasing of all contractual and statutory rights by the tenant. The tenant must therefore account for the fact that in five years the rent will be the open-market rental value of the property as demised at the commencement of the new 20-year lease.

Landlord's present interest as (b) £3,890,529

Landlord's proposed interest

Let proposed rent for 1st 5 years =	£x	
then	£x	
YP 5 years at 7%	4.1002	£4.1002x

Reversion to MRV	£370,000	
YP perp def'd 5 years at 7%	10.1855	£3,768,635 + £4.1002x

present	=	proposed
£3,890,529	=	£3,768,635 + £4.1002x
£121,894	=	4.1002x
£29,722	=	x

Tenant's view:

Tenant's present interest: £870,107

Tenant's proposed interest:

MRV	£370,000
Rent reserved £x	x
Profit rent	£370,000 − x
YP 5 years at 8%/3% 40p	2.5386
	£939,282 − £2.5386x

£870,107	=	£939,282 − 2.5386x
2.5386x	=	£939,282 − £870,107
2.5386x	=	£69,175
x	=	£27,249

Valuing the leasehold on a non tax basis suggests a future rent of £48,518. So although it is shown here that the landlord needs more than the tenant is willing to pay, there is ample room for negotiations provided both parties can see that dual rate adjusted for tax is inappropriate.

(d) Repossession can only be obtained on the following grounds
  (i)   breach of repairing obligation
  (ii) persistent delay or failure to pay rent
  (iii) other substantial breaches of covenants
  (iv) availability of suitable alternative accommodation
  (v)   uneconomic sub-letting
  (vi) demolition or substantial reconstruction
  (vii) required by owner for his own occupation (5-year rule).

The amount of compensation will vary depending upon grounds, but the maximum would be: Loss of security under 1954 Act as specified in the 1990 order is 1 or in the case of 14-year occupation it becomes 2, which on £350,000 is £700,000. Plus Compensation for Improvements (1927) being the lesser of:

  (i)   Net addition to value
  (ii) Reasonable cost of carrying out the improvements at the termination of the tenancy

(i)	Net addition to value	
	Rental value improved	£370,000
	Unimproved	£260,000
		£110,000
	YP perp at 7%	14.2857
		£1,571,427
(ii)	Cost 7 years ago	£300,000
	Increase in costs at say 10% pa	
	× amount £1 for 7 years at 10%	1.9487
		£584,610

Hence maximum compensation might be £700,000 + £584,610.

Chapter 11

GDV

5,000m² (gross) × 85% = 4,250m² net

4,250 at £200m² pa	£850,000	
YP perp at 7%	14.29	£12,146,500
Less purchaser's costs and		
stamp duty at 5.75% (÷ 1.0575)		£325,089
		£11,486,052

Less costs

(a) Building 5,000m² at £400m²	£2,000,000	
(b) Fees @ 12.5%	£250,000	
Total	£2,250,000	
(c) Finance 14% for 1 year on 50% of total	£157,500	
(d) Legal fees (1%) and agent's (1%) on sale		
and promotion	£276,428	
(e) Profit at 15% GDV	£1,821,975	£4,505,903
Total NDV		£6,980,149

Let site value	=	£x
Fees on acquisition + stamp duty	=	4% + 1.75% = 5.75% = 0.0575x
Total debt after 1 year at 15%	=	1.0575x × (1.15)
£6,980,149	=	1.2161x

Site value = £5,739,782 say £5,725,000

This short solution would form the basis for an initial assessment but would need to be checked using a cash flow approach.

Kel and Circle software packages are recommended.

Chapter 13

(a)   IRR = Internal Rate of Return; that is, the discount rate that makes the Net Present Value of a project equal to zero.

(b)   The IRR of an investment before allowing for the incidence of taxation.

(c)   The IRR of a property investment after adjustment for acquisition costs, outgoings but not taxation taking account of current income and reversionary incomes expressed in current value terms.

(d)   The IRR of a property investment after adjustment for acquisition costs, outgoings but not taxation taking account of current income and reversionary incomes expressed in future value terms.

(a)   £1m plus, say, 4% = £1,040,000

(b)   5.77%
$$\frac{60,000}{1,040,000} \times 100$$

(c) 10.58% $\dfrac{110,000}{1,040,000}$ $\times\ 100$

(d) Trial and error try 9.5%
     Term rent   £60,000
     YP 4 years at 9.5%   3.2045   £192,270

     Reversion to   £110,000
     YP perp defd 4 years at 9.5%   7.32183   £805,401
                                                                            £997,671

Allowing for negotiations on sale price, say 9.5%; that is, an equivalent yield of 9.5% (9.15% on £1,040,000).

(e) Rental growth at 5%
     Therefore rent on review =   £110,000
     × Amount of £1 for 4 years at 5%   1.2155
                                                       £133,705

     Value in 4 years' time at 9.5% =   £133,705
                                                        10.5263
                                                     £1,407,421

Therefore possible cash flow adjusted for growth

0 – £1,040,000
1   +   £60,000
2   +   £60,000
3   +   £60,000
4   +   £60,000
                           +   1,407,421

IRR by calculator = 13.048%. This yield might be called an equated yield.

(f) Requirement is an IRR or Target rate of 12%

     Outlay =   £1,040,000
     Income =   £60,000
                                                        £60,000
                                                        £60,000
                                                   £60,000 + £CV

     Term rent   £60,000
     YP 4 years at 12%   3.0373
                                                       £182,238

     Reversion x
     YP perp at 9.5%
     × PV £1 in 4 years at 12%   6.6896
                                               £182,238 + 6.6896x

But
£182.238 + 6.6896x − £1,040,000 = NPV = 0
Therefore
6.6896x = £1,040.000 − £182,238
x = £128,223.21

The rent in Year 4 must have risen to £128,223.
Note: For purists the sale disposal costs should be accounted for and Capital Gains Tax if purchased by a taxpayer.

(i)   £18.223
(ii) £110,000 × Amount of £1 for 4 years at 1%   =   £128,223
       Amount of £1 for 4 years at 1%   =   £128,223 ÷ 110,000
                                                                                  =   1.1656
       (1.1656 − 1) × 100   =   16.56% over 4 years
       and where $(1 + i)4$ =   1.1656
       i = $(\sqrt[4]{1.1656}) - 1$
                                                                                  =   1.0390 − 1
                                                                                  =   0.390 × 100
                                                                                  =   3.9%

Further Reading

The previous editions of this book contained an extensive bibliography of the many journal articles and books which had been referred to in preparing the earlier editions. The further reading listed here is more selective as so much of the material contained in those earlier editions has been subsumed into current practice and theory.

Baum, A. (2002) *Commercial Real Estate Investment*, London, Estates Gazette

Baum, A. and Crosby, N. (1995) *Property Investment Appraisal*, 2nd edn, London, Routledge

Baum, A. and Sams, G. (1991) *Statutory Valuation*, 2nd edn, London, Routledge

Bowcock, P. (1978) *Property Valuation Tables*, London, Macmillan

Bowcock, P. and Bayfield, N. (2003/2004) *Excel for Surveyors and Advanced Excel for Surveyors*, Estates Gazette, London

Byrne, P.J. and Mackmin, D.H. (1975) 'The investment method; an objective approach', *Estates Gazette*, 234: 29.

Crosby, N. (1983) 'The investment method of valuation: a real value approach', *Journal of Valuation no.1*, 341–50, 2: 48–59

Crosby, N. and Goodchild, R. (1992) 'Reversionary freeholds: problems with over-renting,' *Journal of Property Valuation and Investment*, 11: 67–81

Davidson, A.W. (2002) *Parry's Valuation and Investment Tables*, Estates Gazette, 12th edn, London

Dubbin, N. and Sayce, S. (1991) *Property Portfolio Management*, London, Routledge

Fraser, W.D. (1984) *Principles of Property Investment and Pricing*, London, Macmillan and 2nd edn, (1993)

Greenwell, W. (1976) 'A Call for new valuation methods', *Estates Gazette* 238: 481

Harker, N., Nanthakumaran, N. and Rogers, S. (1988) 'Double sinking fund correction methods', University of Aberdeen discussion paper

IVSC (2005) *International Valuation Standards*, International Valuation Standards Committee

Johnson, T., Davies, K., and Shapiro, E. (2000) *Modern Methods of Valuation of Land, Houses and Buildings*, 9th edn, Estates Gazette, London

Marshall, P. (1979) *Donaldsons Investment Tables*, 2nd edn, London, Donaldsons

RICS (2004) *Appraisal and Valuation Standard*, RICS Business Services Ltd, London

RICS (1997) *Commercial Investment Property Valuation Methods: An Information Paper*, RICS Business Services Ltd, London

RICS (1997) *Calculation of Worth: An Information Paper*, RICS Business Services Ltd, London

Rose, J.J. (1977) *Rose's Property Valuation Tables*, Oxford, The Freeland Press
Trott, A. (1980) *Property Valuation Methods: Interim Report*, Polytechnic of the South Bank, RICS

Webliography

www.egi.co.uk — Estates Gazette Interactive
www.rics.org
http://www.isurv.co.uk — RICS Business Services Ltd

Index